RYA
Weather
Handbook

Written by Chris Tibbs

Illustrations by Sarah Selman

Photo Credits: Bluegreen, Rick Tomlinson
Patrick Roach and Helen Tibbs

Our thanks also to: Dundee University
The Met Office and Top Karten

Published by
The Royal Yachting Association
RYA House Ensign Way Hamble
Southampton SO31 4YA

T⌐⌐ ⌐45 345 0400
⌐5 345 0329
@rya.org.uk
⌐.rya.org.uk

Note: While all reasonable care has been taken in the preparation of this
book, the publisher takes no responsibility for the use of the methods or
products or contracts described in the book.

Foreword

Everyone who goes to sea understands the importance of the weather and the necessity of obtaining an accurate forecast.

Good skippers study the weather and plan their trips to take maximum advantage of present and forecast conditions.

This book covers the knowledge required by skippers up to Yachtmaster Ocean standard and provides important background information for the RYA practical and shorebased courses.

Chris Tibbs is a professional meteorologist who has sailed on the Whitbread Round the World race and skippered a Round the World BT Challenge yacht. His clear text provides a wealth of information of value to all skippers from day sailors to Ocean Yachtsmen.

James Stevens FRIN

RYA Training Manager and Chief Examiner

Contents

Contents

Introduction 1

The weather affects our sport like nothing else does. It can turn a pleasant trip into a battle, or vice versa.

Bad weather means different things to different people, depending on the size of their boats, their general level of experience and their perceptions of what is a good sail and what is bad.

We use weather forecasts to help harness the wind to our best advantage and to avoid hazards.

To become a good weather forecaster takes a lifetime of training and experience, but no matter how good we become, computer-generated weather models will always outperform people in predicting large-scale weather patterns. It is better to concentrate on interpreting these forecasts for our area, modifying them for local conditions and, by adding observations, improving on their accuracy over a short period of time. This book is therefore not intended to replace broadcast forecasts but to add to our understanding of the weather.

Weather can be split up into seven variables, all of which are linked, and all of which affect our sailing. For sailors, the most obvious conditions are probably the wind and visibility, but these are linked to precipitation, clouds, temperature, and humidity, and they are all linked to the atmospheric pressure and the general synoptic situation.

None of these variables can be looked at in isolation, and although we are normally interested in only a small geographical area, we have to look at the larger picture before we can refine our forecast to our particular area.

Synoptic scale

At the synoptic scale, meteorologists are concerned with large systems of high and low pressure, and with air masses that may cover many thousands of square miles. This broad-brush picture is often portrayed by a synoptic chart or weather map, showing the meteorological features overlaid onto a geographical base map.

An analysis chart is based on actual observations and measurements at a specific time. From it, meteorologists can produce predicted or forecast charts, which give a general idea of how conditions are likely to change over a wide area.

It is these that are summarized by the general synopsis section of the shipping forecast.

Meso scale

Armed with the synoptic overview, we can look at a smaller area - the area in which we will be sailing over the next few days, perhaps. This is meso-scale meteorology, dealing with areas of a few hundred square miles at a time, or about the size of an individual sea area in the shipping forecast. Meso scale forecasts usually give a

Meteorological terms

Jargon has been kept to a minimum, however some meteorological terms are necessary as they have precise meanings and are well worth remembering. There is a glossary at the back of the book to help with this, and explanations throughout the text where the term is first used.

reasonable indication of the average conditions across the whole area. Bear in mind, though, that there can be a big difference between, for instance, conditions on the west coast of Wales and those on the east coast of Ireland, even though they are both in sea area 'Irish Sea'.

Local scale

What is usually of most interest to the sailor is the local weather, particularly over the relatively short term. Inshore waters forecasts cover this to a certain extent, but they deal with quite long sections of coast: they may not give enough detail to tell us how conditions will change as we round the next headland in half an hour, or to help pick the best spot to anchor within a harbour.

Safety factors

Safety, of course, is paramount, so although there is always the risk of getting caught out by bad weather, an important feature of on-board weather forecasting is that it can help us recognize, well in advance, if a situation is likely to become marginal (or worse!)

This book

This book starts with some theory and a look at the global circulation of air, then concentrates on progressively smaller areas down to micrometeorology, and the circulation of air around clouds.

The behaviour of the wind and weather around coastlines is important, because the land affects the weather in a variety of ways: there is a thermal effect caused by the changing temperature, for instance, as well as a mechanical effect caused by the topography. These can make a significant difference to the wind's direction and strength, but their effects are often so localised that they are not included in the forecast.

Our bad weather is generally from depressions (lows), so a lot of emphasis is placed on how these develop and how they are structured. By understanding this structure and movement a greater understanding of how they will affect us can be reached.

Red sky at night sailor's delight, red sky in the morning sailors take warning. These pictures were taken six hours apart, ahead of an approaching depression. The rising sun is reflected by the invading clouds.

Our *climate* is an average of the weather conditions whilst *weather* is the day-to-day variability. In other words, we look at the climate to decide which boat to buy and what sails to equip it with, but the weather dictates the size of headsail and how many reefs to put in on a particular day.

The single most important factor, which drives the movement of air and thus controls our weather and climate is heat. A factor to remember, and one often forgotten from long ago geography or physics lessons, is that:-*the sun heats the earth's surface, be it land or water, and it is the surface that heats the air above it.*

The sun heats the land quickly and this heat is released into the air. The land also cools quickly at night. The sea is slower to heat and cool, remaining at a more constant temperature throughout the day.

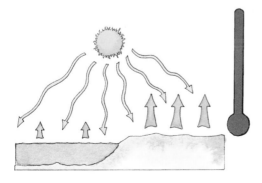

This principle is fundamental in understanding the weather. The air's temperature falls as we go up through the atmosphere to a region called the Tropopause. This is found at an altitude of about 12km in the UK (higher at the Tropics and lower at the Poles). It is in these lowest few kilometres of the atmosphere that all our weather occurs, and in which weather systems grow and decline. We only sail in the very bottom layer, but the vertical structure is important in our understanding of the weather.

Moisture is also important in global circulation. The warmer the air the more moisture it can hold, but as moisture evaporates and condenses latent heat is absorbed and released. Moisture therefore plays its part in moving energy around in the form of heat and latent heat. Surprisingly, moist air is more buoyant than dry, so it also contributes to the vertical movements that are so important in creating weather.

Global circulation

The equatorial regions of the globe are strongly heated by the sun and this heat is transported polewards by global air circulation, and to a lesser extent by ocean currents.

This dispersal of heat towards the Poles helps to regulate the earth's temperature, and as the sea itself has a huge heat capacity it also acts as a regulator, moderating temperatures along coastal regions of the world, whilst extreme temperatures are to be found inland near the centres of large land masses.

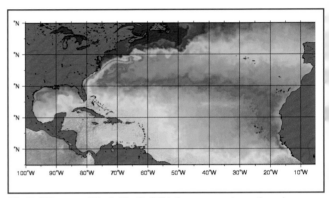

As the surface temperature increases at the equator it heats the air above. (See *Fig 1* below). The warm air rises and expands, which lowers the air pressure at the surface. Air pressure can be thought of as the weight of air vertically above us, so as the air heats, rises and expands, the pressure lowers. The

The Gulf Stream and North Atlantic Drift brings warm water northwards. Red denotes warmest water and blue coldest.

rising air reaches the Tropopause and spreads out, moving towards the Poles. This is a little like a bonfire; the heat and smoke of a bonfire rises to be replaced by cooler air coming in from the side creating circulation, but in this case on a global scale. We see this rising and overturning air on many scales in weather forecasting, from the global to circulation under a convective cloud.

The warm air that is spreading out aloft slowly cools as it moves polewards, causing it to sink and adding to the column of air already there. Thus the air pressure at the surface increases and we get an area of high pressure. The surface air under the higher pressure is moved away – squeezed out - and heads towards low pressure.

Fig 1 Theoretical circulation between the equator and the Poles.

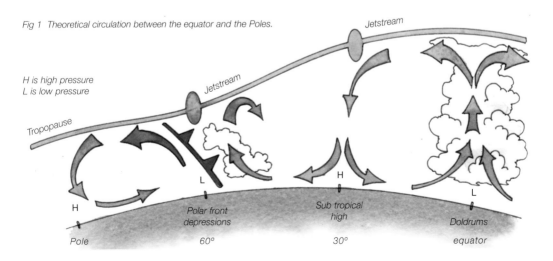

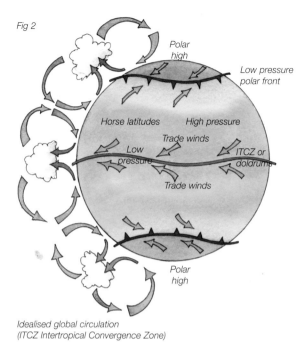

Fig 2

Polar high

Low pressure polar front

Horse latitudes   High pressure

Trade winds

Low pressure   ITCZ or doldrums

Trade winds

Polar high

Idealised global circulation
(ITCZ Intertropical Convergence Zone)

Two more fundamentals of weather and systems are illustrated here (*Fig 2*); surface air moves from high to low pressure, and air in low pressure rises. The opposite happens in high pressure, i.e. air descends and moves away from the centre at the surface.

This generates the world's main wind patterns and would all be very neat and uniform if it were not for the uneven distribution of land, and the effect the spinning earth has on our lives.

The spinning earth causes the wind around depressions in the Northern Hemisphere to circulate anticlockwise and to circulate clockwise around high pressure. The directions are reversed in the Southern Hemisphere; more about this in Chapter 16.

Air masses

Because the air spends long periods of time over land or sea it takes on certain characteristics, such as, temperature and humidity. Air over cold land, for instance, cools and is relatively dry. Over a warm sea however it becomes warm and moist. We have already seen that the warmer the air, the more moisture it can hold before it becomes saturated. This is particularly relevant for sailors around western coasts where most of the weather brings air that has spent a long time over the oceans.

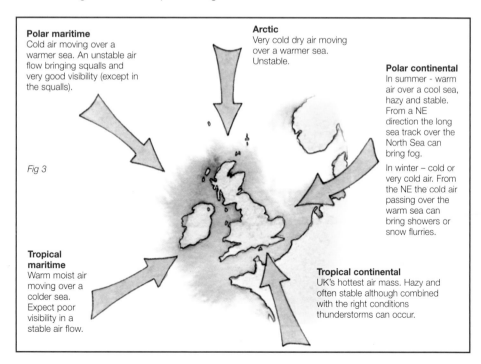

Polar maritime
Cold air moving over a warmer sea. An unstable air flow bringing squalls and very good visibility (except in the squalls).

Arctic
Very cold dry air moving over a warmer sea. Unstable.

Polar continental
In summer - warm air over a cool sea, hazy and stable. From a NE direction the long sea track over the North Sea can bring fog.

In winter – cold or very cold air. From the NE the cold air passing over the warm sea can bring showers or snow flurries.

Fig 3

Tropical maritime
Warm moist air moving over a colder sea. Expect poor visibility in a stable air flow.

Tropical continental
UK's hottest air mass. Hazy and often stable although combined with the right conditions thunderstorms can occur.

Stability

The air from any particular region has very small differences in temperature or humidity over an area spanning hundreds or even thousands of kilometres. This is important for us as these characteristics define what the weather conditions are likely to be within that air mass.

The diagram (*Fig 3* on page 13) shows where different air masses come from for Northwest Europe and gives the main characteristics.

The origin of the air gives it its characteristics and the track that the air takes will modify the characteristics. As a rule, the temperature of the surface that the air is moving over will either heat or cool the air mass from below. This will make it either more or less stable. See Chapter 10 on stability.

Unstable mass after a cold front has passed.

The importance of where the air has spent its time recently becomes obvious when looking at depressions. In the warm sector, the moist tropical maritime air brings low cloud and drizzle to the West Country and in the Spring, fog forms over the cool sea which will persist until the air mass changes.

Air masses move, but when they meet each other they don't usually mix very readily. This means that where two air masses meet, there are likely to be sharp changes in temperature and humidity, described as fronts.

Stable air mass with fog and mist in tropical maritime air.

The most dramatic change in air mass that we see is when a cold front passes. After the drizzle then heavy rain, the sky clears giving good visibility, a noticeable drop in temperature, and at times squally showers.

Unstable air	Stable air flow
Big vertical movement in the atmosphere giving heaped clouds and gusty conditions.	There is little vertical movement in the atmosphere, giving flat layers of cloud.
Typical of cold air moving over a warm surface.	Typical of warm air moving over a cold surface.

Weather charts **3**

Representing the weather on a flat piece of paper is similar to representing the land on an Ordnance Survey map.

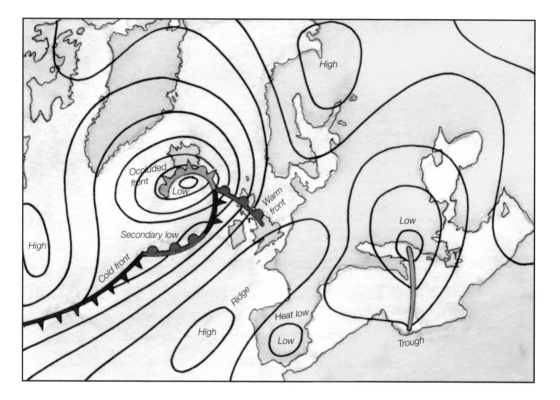

Synoptic weather charts and what they mean

The main variable on a weather chart is the surface pressure. This is measured in millibars (mb) or hectopascals (hpa) and is represented on a synoptic chart by lines called isobars.

Like contours on a land map joining places of the same height, isobars join places recording the same pressure. Low pressure is like a valley, and high pressure is like hills and mountains. Even some of the jargon is the same, with ridges and cols around the mountains of the high whilst troughs extend from lows. The diagram above shows a weather chart with some common features labelled.

The analogy can be taken one stage further with the wind moving from high to low pressure just as gravity would move an object on the land. Here, though, the similarities end as the rotation of the earth takes control and bends the surface wind to make it follow the isobars with only a few degrees of outflow from highs, and flow inward to lows (see *Fig 4* overleaf).

On a land map, the closer together the contours the steeper the slope of the hill. On a weather chart, the closer together the isobars, the greater the pressure gradient and the

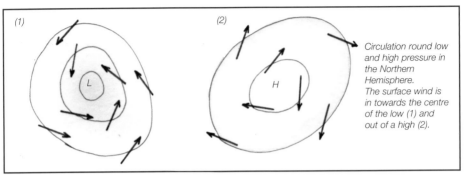

(1)

(2)

L

H

Circulation round low
and high pressure in
the Northern
Hemisphere.
The surface wind is
in towards the centre
of the low (1) and
out of a high (2).

Fig 4

stronger the wind. This is important when interpreting weather maps as there is a direct relationship between the pressure gradient and the wind speed. We will look at measuring the wind speed from a chart in Chapter 8 and see how to adjust it to give our sailing wind.

The boundary layer

We sail in what is known meteorologically as the boundary layer, i.e. the air closest to the earth's surface and where most of the interaction between the surface and atmosphere occurs.

The surface has a profound effect on the boundary layer so what happens close to the surface can be quite different from what happens much higher up (see Fig 5). There is no strict height definition for the boundary layer and its extent varies with conditions; for working purposes we can use 600 metres (about 2,000ft).

Fig 5   We sail in the boundary layer where drag and turbulence are found. The more obstructions to the wind there are, the rougher the surface is said to be.

Although the weather maps we use are for the surface, in reality they are giving us information about the wind at the top of the boundary layer. This has to be modified to be of real use to us.

Above the boundary layer, the wind follows the isobars and runs parallel to them. It is also convenient to measure the wind's strength by measuring how close together the isobars are and hence the gradient of the wind. The standard isobar spacing is 4mb – but do check, as some charts found on the internet will have 5mb spacing and others 2mb.

Isobars

The closer together the isobars the steeper the gradient and the stronger the wind.

Symbols found on a synoptic chart

Fronts

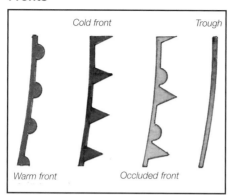

Fig 6

Station circles

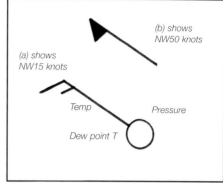

Fig 7

Some charts may have station circles marked showing the actual data recorded at the weather stations concerned.

The most useful information for sailors is in the form of a wind arrow attached to the circle, where each full barb represents 10 knots, and a half barb is 5 knots. The wind blows from the barbs to the centre of the circle, so in our example the wind is blowing from the northwest at 15 knots (*Fig 7a*). If the barbs are replaced by a filled-in triangle (*Fig 7b*) it represents 50 knots. Wind arrows may be shown on their own without a circle.

Other common symbols and their meanings

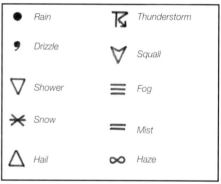

Fig 8

Other information may be found around a station circle in the positions shown and the circle itself may be filled-in to show the extent of the cloud cover (*Fig 9*).

Station circles filled in as shown denote cloud cover

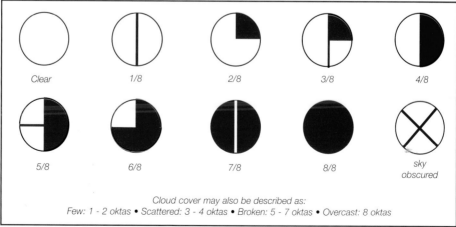

Cloud cover may also be described as:
Few: 1 - 2 oktas • Scattered: 3 - 4 oktas • Broken: 5 - 7 oktas • Overcast: 8 oktas

Fig 9

Low pressure and the polar front

The weather that dominates most mid-latitude regions both north and south of the equator is caused by the procession of low pressure systems that arrive from the west.

There are preferred locations where cyclogenesis (the creation of depressions) most often occur; for Western Europe this is the North Atlantic. The depressions then sweep in across, or close to, the UK; few forecasts go by without the mention of at least one 'low'.

Fig 10   The polar front in equilibrium.

The majority of these depressions form along what are known as the polar fronts (as shown in the global circulation diagram on page 13). This is where the air circulating around the subtropical high meets the colder air flowing in the opposite direction.

These two air masses meet but do not mix and a small disturbance on this polar front causes a depression to form. The spinning earth forces this air to rotate anticlockwise in the Northern Hemisphere and a low-pressure system is born (see Chapter 5).

The air masses retain their original characteristics and do not mix with each other so there is a distinct dividing line between them, called a front, like the front line between opposing armies (Fig 10). Interestingly, this idea of the weather systems was devised by the Norwegians during the First World War. The concept of armies fighting along fronts was easily transferred to the weather where large temperature and humidity differences 'fight it out'. The model devised by the Norwegians is still in use today and although sometimes a long way from reality it is useful to illustrate what happens. It must, however, be used with some caution as the weather does not always conform and no two depressions are ever identical.

The polar front jet stream, often referred to as the jet stream, is situated between warm and cold air masses and follows a generally west to east direction.

Clouds 4

Before we take a closer look at depressions it is worthwhile taking time to study clouds. On all but the clearest days there are clouds above our heads. There is a message in these clouds if we only can read it. With a bit of practise, we can begin to understand why the clouds are there, what they are telling us, and how to use the message to our advantage.

Whilst the wind in our faces is a indication of horizontal air movement at the surface, the clouds give a visual indication of what is happening above us. They show the vertical profile of the weather, giving a three-dimensional picture of the atmosphere's movement.

The clouds' tracks across the sky show the direction of upper level winds. Good examples of this are cirrus clouds (very high and wispy clouds sometimes referred to as mares' tails) invading the sky from the west whilst the surface wind is southerly. These are a sure sign, and usually the first, of an approaching depression. Low clouds or scud (small fragmented low cloud) moving quickly may indicate a strong wind to come or, that the wind is strong out at sea.

We are lucky, these days, to be able to access satellite images to see the top of the clouds as well as looking upward to see the bottom. With a bit of knowledge we can then take an educated guess at what is happening in-between.

On a synoptic scale, clouds will tell us where we are within the system, and on the local scale, indicate gustiness and wind shifts.

Synoptic charts will give us an idea of what should be up there, but often there is such a jumble of clouds at different levels and of differing shapes and colours, that it is tempting to give up and just say "it is cloudy". Although there are tens, if not hundreds, of cloud classifications and sub-divisions to describe what we see, clouds can also be divided into much simpler, general groups based on appearance and height.

The main three are **cumulus** (a heaped or lumpy cloud), **stratus** (layers of flat clouds) and **cirrus** (high wispy clouds). By adding a height characteristic of **cirro** for high clouds and **alto** for medium levels we have our main classifications. There is just one extra addition and that is **nimbus** for when the cloud is rain (or snow) bearing - whether it is actually raining, or just looks as if it may do. Cumulonimbus, for instance, is a heaped rain cloud while nimbostratus means it is raining from thick layer clouds.

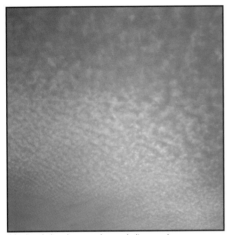

Mackerel sky-cirrocumulus and altocomulus

Altostratus

Types of Cloud

	Name of cloud	Abbreviation	Height band
High layer	Cirrus – wispy, hair like	Ci	Above 16,500ft (5,000m)
	Cirrostratus – fine layer, possible halo	Cs	
	Cirrocumulus – lumpy layer	Cc	
Medium layer	Altostratus – thin layer	As	6,500ft – 16,500ft (2,000-5,000m)
	Altocumulus – lumpy layer	Ac	
Low layer	Stratus – uniform layer	St	Surface to 6,500ft (0-2,000m)
	Stratocumulus – undulating or lumpy layer.	Sc	
	Nimbostratus – thick rain bearing layer	Ns	
	Cumulus	Cu	From low level to tropopause - approx 38,000ft (12km)
	Cumulonimbus	Cb	

Note: *Cloud heights are traditionally given in feet and reports will usually be given in feet or 000s of feet.*
Approximate heights are given as a guide only; clouds may extend through the bands and it is impossible to measure the height accurately by eye. In met. reports, cloud cover is given in oktas or eighths with 8/8 signifying an overcast sky (see page 17).

Judging the height of clouds is difficult even for a trained observer, but it becomes easier with practise, and especially when they are seen in context of a large scale system.

Cumulus tends to be hard to define by height; although the base is low, the vertical extent may be 30,000 feet (as in the case of cumulonimbus clouds). The classification is therefore the height of the base.

> Clouds are formed by air lifting and cooling until the dew point is reached when the water condenses and becomes visible.

Extensive layer cloud

Extensive layer cloud forms when there is large-scale ascent; mainly in depressions and around fronts. This explains the cloud structure ahead of a warm front when warm, moist air is lifted over colder air (see anatomy of a low on pages 30/31). Large-scale or mass ascent also occurs when air is lifted over the land. This is called orographic uplift and is often visible when sailing near high coasts.

Heaped or cumulus clouds

Heaped or cumulus clouds are formed by convection. Air and moisture are heated and rise from the surface forming a local thermal. To remain in equilibrium the air that rises must be replaced, giving a continuous local circulation (*Fig11*).

This produces individual clouds that tend to form lanes across the sky but there are clear patches where the cold air returns to the surface. Cumulus and their big brothers, cumulonimbus clouds, by their very nature generate this local air circulation that is so important to us when sailing (see Thunderstorms on page 79).

Fig 11   Cumulus cloud and its circulation

Stability

Of all the messages that the clouds are telling us, one of the most important for sailing is about stability of the atmosphere (see page 14). We all know that hot air rises. If the rising warm air is lifted into relatively cooler air it will keep on rising, and clouds of great vertical extent are formed – towering cumulus or cumulonimbus. However, if the air above is relatively warm then there will be little vertical movement and layer clouds will develop.

Therefore, big cumulus clouds are an indication of instability and we can expect gusty conditions, showers and possibly squalls. Layer clouds show stability giving fewer and lower gusts, and any increases in the wind are likely to be gradual.

The stability will also tell us where we are within a system and the likely air mass that we are in, allowing us to make a better judgment in both the short and long term.

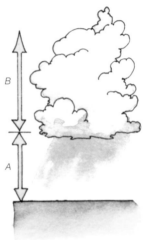

Marking landfalls

Cumulus clouds are often triggered by land or hills in the path of the air flow, and there is often a cumulus cloud sitting on top of a high island.

If the airflow is stable, clouds may also mark a landfall with a banner cloud over the high land. In both cases the cloud over the land can be seen well before the land itself and has been a great indicator for mariners since boats first sailed beyond the sight of land.

Banner cloud

Rainfall

Fair weather clouds.
The height of the base above the sea is greater than the thickness of the cloud.

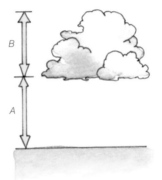

Showers are possible when the height above the sea becomes about equal to the thickness of the cloud. A = B.

Showers accompanied by gusts are probable, once the thickness of the cloud is greater than the height of the cloud base above the sea. B is greater than A.

Apart from making us wet and correspondingly miserable, rain has significant effects on sailing and can tell us a lot about the weather. Heavy rain seriously reduces visibility and rain squalls bring strong gusts. The intensity of the rain is also an indication of where we are situated within a depression – more on this when we look at the anatomy of a low and its life cycle (pages 25/27).

An interesting point, and one that helps in understanding forecasts, is that irrespective of how long the rain lasts, showers only come from convective clouds i.e. cumulus and cumulonimbus.

For short bursts of rain from layer clouds, expected to last less than one hour, the term 'intermittent' rain is used (and 'continuous' if it were expected to last for over an hour). It may seem pedantic but it does help to describe the rain, and gives a good indication of where we might be within a depression.

Significant points

Clouds show the stability of the atmosphere.

They show where we are in a system.

They can be the first indication of approaching depressions.

They indicate short-term changes in the wind.

They indicate rain and reduced visibility.

Additional messages

Clouds are not just random events and any change in the clouds will mean something, although exactly what is not always apparent. Often, near land, it will indicate thermal or mechanical activity in the atmosphere (see Chapter 10).

Lack of cloud signifies an area where the air is descending. It may be small-scale around cumulus clouds or large-scale found with high pressure. With any descent the air warms as it falls hence any moisture tends to evaporate rather than condense.

High pressure in the Mediterranean.

Extra theory for the interested

The air cools with height at about 6.5°C/1,000m. This is called the lapse rate. The atmosphere is rarely uniform though, and as we move vertically upwards through it there are bands of warmer and colder air. If we imagine a small parcel of air at the surface and heat it up a little, it will naturally rise but as it does so it will also cool at approximately the lapse rate. It is the temperature of the air which surrounds our parcel that determines whether it continues to rise or stabilises. It is therefore the temperature profile of the atmosphere that is important and the clouds are an indication of this profile.

The majority of low pressure systems that we see in NW Europe develop along the polar front, out in the Atlantic. Plenty of heat and moisture are provided by the air circulating around the Azores High, augmented by the Gulf Stream, whilst cold dry air arrives from the north.

The life cycle of a depression

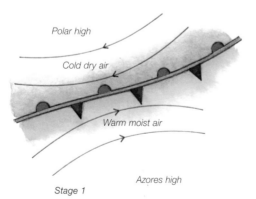

Stage 1

Stage 1. The polar front in equilibrium.

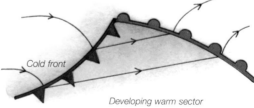

Stage 2

Stage 2. A disturbance in the upper atmosphere causes a wave to form on the polar front. The tongue of warm air in an area of generally cooler air causes a local reduction in pressure. This embryonic low pressure develops its own circulation, which helps make the wave even more pronounced.

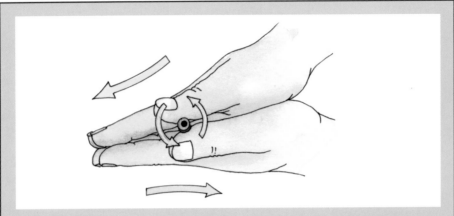

Hold a pencil between your hands and move in the direction of the arrows in the diagram. The pencil is like a disturbance on the Polar front and will rotate like a fledgling depression.

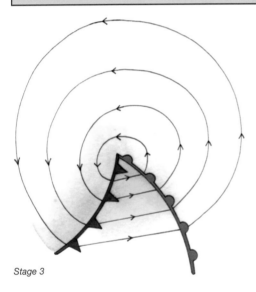

Stage 3

Stage 3. A full-blown depression. The arrows show the wind direction for a depression in the Northern Hemisphere. The surface wind is angled about 15° inwards from the isobars towards the centre of the low.

Cold fronts and warm fronts are the leading edges of the air masses they describe, hence the warm front in a depression is the leading edge of the warm sector with the warm air mass following on behind. The cold front is the leading edge of the cold sector.

In decline

By the time the majority of depressions reach European shores, they have reached their maturity and are in decline.

Over time, because the cold front is moving faster than the warm, it catches up, squeezing the warmer air in the warm sector off the surface as an occluded front is formed. Occluded fronts are a mixture of warm and cold front features and can be divided into warm and cold occlusions depending on the relative temperature of the air that is being caught up. When looking at a weather map, if the occlusion is an extension of the cold front then it is a cold occlusion and if an extension of the warm front, a warm occlusion. The majority of occlusions in the UK are of the cold type (around 70% especially in summer) but in practical terms the only real difference is the amount of rain and the structure of the cloud to be found there.

On the weather chart, occluded fronts are shown as a mixture of the warm and cold front symbols and if drawn in colour are purple.

Stage 4. The low is in decline and shows a bent-back occlusion and a secondary low developing on the trailing cold front. This is not uncommon, and families of depressions can cross the Atlantic, each one tracking a little further south until the thread is broken and a new polar front develops to the north.

Another feature of an occluded front is the way in which it can become bent back around the centre of the low. Passing over and around the centre it produces a big band of cloud and rain.

As the low is generally filling and slowing down this band of heavy cloud can bring persistent rain that is reluctant to clear. Sailing under this band of cloud can be frustrating as the wind dies and often becomes variable in direction.

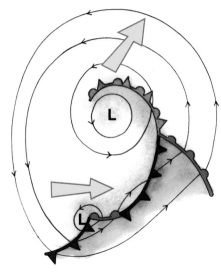

Stage 4

Occlusions

Once occlusions begin to appear it is a sure sign that the depression is reaching the end of its life cycle. Although the low may deepen for a further 6 - 12 hours after the fronts start occluding, it will then start decaying and filling. As a rule of thumb the low will also start tracking further to the left of where it had been heading, that is northwards if the low is following its usual path across the Atlantic from the west to east.

As the low decays still further, the fronts will be seen to become detached from the low and the centre expands to cover a greater area. This is the depression going into decline, filling and becoming a less significant feature on the weather charts.

Non-frontal depressions

Although the majority of depressions reaching our shores are from the Atlantic and are created along the polar front, there are other forms of low pressure that are also common throughout Europe and the rest of the world; these can be grouped together under the heading of non-frontal depressions.

The first we shall look at are lee depressions and lee troughing. These are found downwind of mountains and high ground.

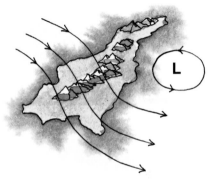

Fig 12 The flow of wind over a mountain range produces lee troughing and lee depressions.

Lee depressions

As air flows over mountains and high land an area of low pressure develops in their lee (*Fig 12*). It may develop as a trough or generate its own circulation and form a full-on depression. There are many well-known examples where this happens regularly. The Gulf of Genoa in the Mediterranean is a place where low pressure systems develop. A northerly flow is blocked by the Alps, to be released by the wind rushing down the Rhône Valley. This not only produces a strong Mistral but an area of low pressure develops in the Gulf of Genoa. During the winter months a weak polar front can occur here, with cold dry air from the north meeting the warm moist air over the Mediterranean Sea.

Further afield the mountains of New Zealand block the prevailing westerly wind; strong winds develop around their edges and a low forms in their lee. This will be discussed further in the chapter on dangers, as these strong winds may not show up on the sea level pressure charts although they should be given in the local forecasts. Sailing in the lee of mountains may therefore not afford the shelter that you expect.

Heat lows

During the summer months the heating of the land creates semi-permanent low pressure areas. On a global scale the bigger and hotter the land mass the more pronounced the low pressure. These heat lows affect the winds around them – in Europe the summer heat low over Spain (*Fig 13*) becomes a semi-permanent feature and drives the Portuguese trade winds. These winds blow down the coast of Portugal making it difficult at times to make progress sailing north against them.

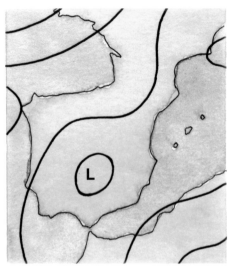

Fig 13

Further east in the Mediterranean the Meltemi is driven by the heat low over Turkey and Asia. These lows are non-frontal and the wind around them increases during the day, and eases at night as the land heats and cools.

Polar lows

The final form of depression is the polar low; caused when very cold Arctic air moves south during the winter (in the Northern Hemisphere), over a warmer sea. A small intense depression can form bringing gale force winds and heavy snow falls. The lows show up on satellite pictures as relatively small features, gaining their energy from the temperature contrast between the relatively warm sea and the cold air. Moving over land, the source of energy is switched off and they quickly decline.

Buys Ballot's Law

The wind around the depression spins anti-clockwise in the Northern Hemisphere and can be remembered by using Buys Ballot's law, enunciated in 1857. It states that when standing with your back to the wind, holding out your left arm will point to the centre of low pressure; the opposite holds true in the Southern Hemisphere.

Depressions in practice 6

Depressions tend to dominate our weather. They are also the features most likely to bring dangerous conditions so they are also the ones we worry about the most.

A practical look at depressions

There are many old sailors' rhymes that describe the weather; the majority are connected to warnings about depressions and bad weather.

We have seen that lows usually develop well offshore, so by the time a low reaches our coastal waters, it is likely to be an established feature already at the mature stage or in decline (*Fig 14*).

A Sailors Rhyme

With a low and falling glass, soundly sleeps a careless ass - only when 'tis high and rising soundly sleeps a careful wise one.

One of my favourite rhymes warning of the danger as a depression approaches.

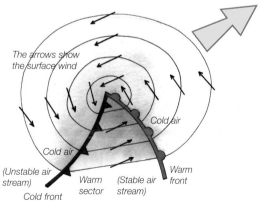

Fig 14 A typical Northern Hemisphere depression. The arrows show the surface wind.

The anatomy of a low

Cumulus cloud after the cold front has passed

Stratus in the warm sector

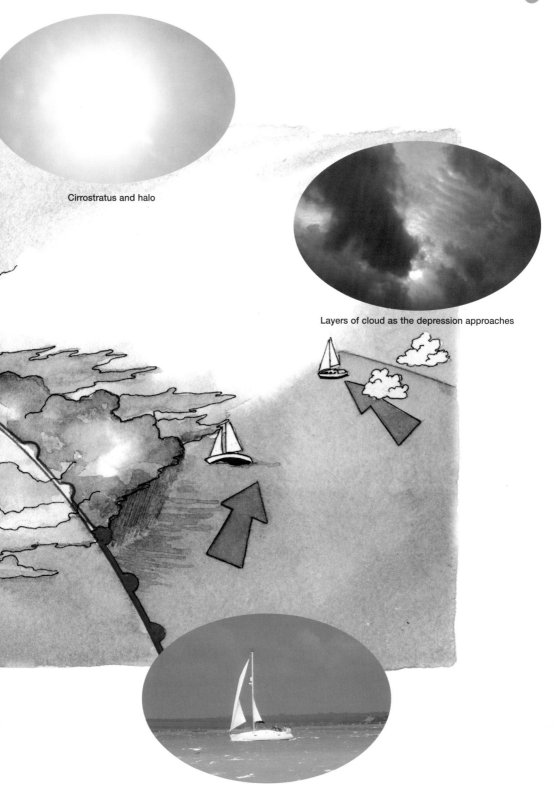

Cirrostratus and halo

Layers of cloud as the depression approaches

Lowering cloud and rain ahead of the warm front

Ahead of the warm front

Initially, we will look at what conditions are likely to be met if sailing south of the track of a depression. This is not always the case, particularly if in Scotland or more northerly latitudes. So we will look at what happens if the depression is likely to come directly overhead, or pass south of us, at the end of the chapter.

A pleasant day's sailing with a few cumulus clouds around may be coming to an end and as the cumulus clouds become fewer it is noticeable that there are high wispy clouds invading from the west. This sheet of cloud is usually the first indication that a depression is on its way. The wispy clouds are made up of ice crystals and are known as cirrus (commonly called mares' tails). The speed at which the cirrus arrives indicates the speed of movement of the low. 'Hooks' (that look like fish hooks), in the clouds show that they are close to the jet stream, where winds in excess of 100 knots may be found at about 30,000ft.

As the cirrus invades, any cumulus clouds around are likely to decline in size and frequency, possibly making it hard to relate the fine current weather with a forecast of stronger winds and rain to follow. At this stage, the low pressure is typically still 600 miles away.

This gives rise to the old saying - *Backing winds and mares' tails make tall ships carry low sails.*

Early indications of low pressure

High cirrus clouds.

A backing of the wind to the south.

The barometer begins to fall.

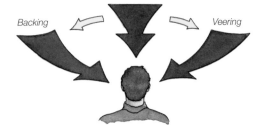

Backing

Veering

Backing – a change in the wind direction in an anticlockwise direction, e.g. from W to S.

Veering - is the opposite to backing. A change in the wind direction in a clockwise rotation e.g. from west to north.

The terms hold true in either hemisphere.

The cirrus, however, is only the thin end of the wedge and provides the tell tale sign of the warm front that divides the cool air from warm air following the front. This warm sector air is riding up and over the colder air along a slope of about 1 in 150. The front is still a long way away but as it advances, its height above our heads is reducing.

We have already seen that clouds form where there is a large-scale ascent of air (see Chapter 4 on clouds). The warm air riding up over the cold air gives us this large-scale ascent. As the warm air rises it cools, hence the clouds are formed.

As the depression moves closer, the cirrus cloud thickens, to form a whitish, milky veil or layer of cloud called cirrostratus. The sun appears hazy through the cloud and often there is a halo around it. This halo is a luminous ring caused by the sunlight being refracted through ice crystals. Although it can occasionally be seen at other times it is usually an early indicator of the approaching warm front and depression.

This phenomenon is not confined to the sun - it can also be seen around the Moon, where it can be very clear and rather beautiful even though it is an early indicator of deteriorating weather. A break in the halo is often said to indicate the direction the bad weather is coming from, although in practice it is usually caused by the low altitude of the sun or moon.

> The length of time between the cirrus arriving to losing the sun, is usually about equal to the time from losing the sun to rain falling.

By now the wind should have backed to the south or southeast and the barometer will have started a slow steady fall. The speed with which the barometer falls indicates the depth of the depression and how quickly it is moving, whilst how far the wind swings to the south indicates where the centre of low pressure is likely to be. (Remember Buys Ballot's Law on page 28).

As the cloud thickens the halo disappears, and the sun looks as if it is shining through ground or frosted glass. This is altostratus - sheets of thickening lowering cloud giving a gloomy appearance that could best be described as blue-grey 'mudflats' in the sky.

The approaching warm front lowers the clouds until it begins to rain. At first, the rain may appear to hang and not reach the surface, as scud (small ragged fragments of low cloud) rushes across the sky. Although the rain may start off fairly light, it will increase in intensity as the front approaches. A characteristic of the rain is that once it has started it is persistent, varying in intensity and becoming generally heavier as the layers of cloud thicken.

If the barometer falls rapidly

The faster the barometer falls, the stronger the wind. As a rule of thumb, a 6mb drop in three hours indicates a Force 6, whilst 8mb in three hours indicates a Force 8. Any faster sustained fall would indicate an even stronger blow.

The approaching warm front

The cloud thickens and lowers until it starts raining; the rain can be heavy and persistent.

The barometer falls steadily. The faster the fall, the stronger the wind.

The wind backs to the south and increases.

Good visibility becomes poor in the rain.

Speed of the depression

Timing is always difficult in meteorology and becomes more so as the depressions move over the land and the progress of the fronts may be interrupted.

The speed at which the warm front travels is a little slower than that of the surface wind speed behind it. This speed can be calculated more accurately from a weather map, or a series of charts, if they are available (see Chapter 11). Listening to the forecast, however, will give an indication of where the front is, and its speed of movement can be estimated from the predicted wind behind the front.

Sitting in harbour waiting for the front to clear can be frustrating. The forecast wind may not be all that high, but few people enjoy sailing in the rain. Here the local land forecasts can help to plan the day as they are much more interested in the rain than wind.

The local land forecast will concentrate on where the rain is and when it is likely to clear. By correlating this with the sea area forecast and our knowledge of the behaviour of the warm front, a good estimate can be made of the time of the wind swing and the clearing of the rain.

Cold fronts travel at approximately the same speed as the surface wind behind them. As this is usually the strongest wind in a depression, the cold front moves faster than the warm front, catching it up over time.

It is useful to remember that the depression itself will usually follow a course that is parallel to the isobars in the warm sector.

As the warm front arrives

A lightening of the sky to windward will herald the front, the heavy rain will ease, and a wind shift from south to southwest is to be expected. Shortly ahead of the front, the wind may actually back a few degrees, although this is often missed unless you keep a very close weather eye. How much the wind veers on the front varies every time - it may only be small, or it may be 40 degrees or more.

As the warm front goes by

The barometer should abruptly stop its downward plunge and level off, becoming steady as we enter the warm sector. One of the less instantly recognisable changes is an increase

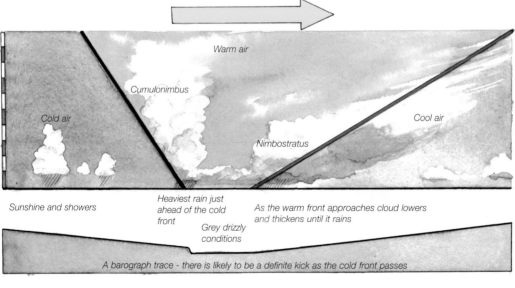

Warm air

Cumulonimbus

Cold air

Cool air

Nimbostratus

Sunshine and showers

Heaviest rain just ahead of the cold front

Grey drizzly conditions

As the warm front approaches cloud lowers and thickens until it rains

A barograph trace - there is likely to be a definite kick as the cold front passes

The barometer

The warm front arrives and passes through

The sky will lighten on the windward horizon.

There will be a break in the rain.

The wind will veer from S to SW, but it may back a few degrees immediately ahead of the front.

A steadying in the barometer's fall.

A rise in the air temperature.

Not all the signs will be apparent on every warm front.

What to expect in the warm sector

A steady wind in direction and strength, usually from the SW.

Steady pressure.

Low cloud with some breaks.

Warmer than before although it may be damp and drizzly.

Moderate or poor visibility with fog possible.

in temperature - but unless you are constantly monitoring with a thermometer, it is only after the rain has stopped that this becomes a recognisable feature.

Warm sector weather

The air in the warm sector is a tropical maritime air mass, meaning that it has spent its recent past life over warm seas and has acquired some of their characteristics (see air mass theory on page 13). The relative temperature and humidity of the air are high - hence the name 'warm sector'. This may give a muggy warm feeling and reduce the visibility in the warm sector.

As the air mass is cooled by the colder sea it is now passing over, it becomes increasingly stable; there will be less vertical movement in the air and the wind should be steady in both direction and strength. The barometer should also have levelled off. If the barometer is still falling, then the depression is still deepening or the centre of the low is moving closer, and we can expect the wind to increase.

The cloud in the warm sector will bring drizzle or intermittent light rain and the visibility is usually poor with mist or fog likely. This depends mainly on the surface temperature of the sea; if it is cold compared to the air temperature, the lowest level of air will cool to below its dew point causing fog to appear. The speed of the wind will play its part as the stronger it blows the greater the mechanical turbulence and mixing, lifting the fog to a layer of low thick cloud. 15 knots, the top end of a Force 4, is usually the upper limit of wind to still have fog around.

The cloud in the warm sector varies enormously depending on the recent track of the depression. On western coasts, there is often rain and drizzle with poor visibility but on the east coast, the air will have dried out whilst crossing the land. This can give quite different conditions between east and west coast warm sectors.

The further from the centre of the low we are, the less cloud and rain we can expect. Although a depression passing over the north of the UK will produce the expected weather along the fronts, the south coast is likely to see the clouds breaking and may even get a reasonable amount of sunshine during the warm sector, whilst close to the point where warm and cold front meet, the cloud will be considerably thicker and the rain more persistent.

As the air mass is stable, convection is restricted and any breaks in the cloud will show that the upper cloud has all but gone.

On weather charts, the warm sector is recognisable by the almost parallel straight lines of the isobars between the warm and cold fronts. This gives a steady wind in both direction and strength, the actual strength depending on how close the isobars are together.

The cold front

The first sign of the approaching cold front is a thickening of the cloud. If there are any breaks in the cloud, then it will be seen reaching to much higher levels. As the cloud

thickens and the rain starts, or intensifies, then the cold front is on its way. As it is not always possible to see the thicker cloud, the increase in rain is often the first sign of the cold front. There will not only be more of it, but the rain drops will be much larger. In fact, the whole characteristic of the rain changes.

The cold front arrives

As the cold front arrives the rain intensifies and the wind becomes gustier. This is because it is coming from clouds that extend vertically up to 20 or 30,000 feet and produce large-sized rain drops. This band of heavy rain is usually around 50 miles wide although it can be double this, or it may be just a short shower.

Embedded in amongst the cloud can be thunderclouds generating squalls and thunderstorms as the front goes through. The heavy rain can be taken as both a good and bad sign; bad because it can be quite unpleasant sailing in squally conditions and good because the cold front is nearly upon us and the weather will improve. A note of caution, however; as the cold front passes, the wind is likely to increase in strength and swing quickly, sometimes violently, to the northwest becoming gusty and strong. Once the rain has stopped, visibility will greatly improve in the colder, clear air.

As the rain eases, a brighter band will be seen in the north west. This band will move rapidly towards us as the front passes through (see page 79).

Within a normal depression one of the fronts is stronger than the other. In the UK, it is usually the cold front that packs the punch but this is not always the case. Looking at the weather charts, the isobars are usually the tightest together behind the cold front signifying where the strongest wind will be.

Cold clear air with great visibility

The wind behind the front will change. This is because the air mass has changed; the air is colder, usually coming from the northwest and probably not all that long ago it was over Canada or even the North Pole.

Any dirt or pollution will have been washed out and the amount of moisture suspended in the air is low because of the colder nature of the air. This will therefore produce different clouds form from those found in the warm sector. In fact, immediately behind the cold front there is often a band with no clouds at all. This will be followed by building cumulus and cumulonimbus clouds that, when big enough, will produce showers of building intensity, sometimes accompanied by hail or thunder. This is unstable air and will generate large convective clouds and showers. The instability will bring strong gusts.

Unstable air

Wind

As the cold front approaches, the wind will increase a little and there is a good chance that it will back a few degrees, (i.e. swing a few degrees in an anticlockwise direction). The often-torrential rain may disguise this to a certain extent, but as each front is different it may not back at all.

The wind will then become more blustery and veer to the northwest as the front goes through. Sometimes, but not often, we get a full 90° swing in the wind; most other times it will be much less. Weather maps and forecasts will help give an indication on how much the wind will swing and how strong it will get. After the front, the nature of the wind will have changed and we are now in a blustery, squally air stream. The wind behind the cold front is usually the strongest found in the depression with the isobars packed closely together.

As the cold front passes

The wind will back a few degrees then increase and veer. The change is likely to be in a squall and can be violent, sometimes with thunderstorms.

Heavy rain ahead of the front will give way to a clearing sky.

The pressure will start to rise, often with a kick.

Visibility will be poor in the rain, becoming very good.

Sunshine and showers

'Sunshine and showers' describes the weather behind the cold front. Some of these showers can be heavy: the large towering cumulus clouds following the front often fall into lanes and the showers they produce are not randomly scattered but follow down these lanes.

At other times, the clouds will form a band across the flow of the wind producing a trough or mini front - shown as a black line on weather charts. This trough will behave as a little cold front bringing a band of heavy rain, backing the wind a few degrees ahead of it, and veering on passing.

Whilst this is going on, the barometer, which will have been steady or in slow decline during the warm sector, will give a sudden jump or 'kick' as the front goes through. It will then start a steady and sometimes spectacular rise. A fast rise, like a fall, can bring strong winds leading to the saying: *A quick rise after low predicts a stronger blow*.

Occluded fronts

In the section on the life cycle of a depression we saw that by the time depressions reach European shores the majority will be in decline. The central pressure will be filling and the cold front will be catching the warm front, squeezing the warm sector and lifting the warmer air aloft. Where the fronts join they are said to occlude.

Where the fronts have occluded, the mixture of warm front and cold front clouds is likely to produce heavy rain, and a reluctance for the front to clear. Over land, the fronts can stall: such slow-moving occlusions and the prolonged heavy rain they produce are often the cause of flooding ashore.

The majority of occlusions are cold occlusions in which the air behind the cold front is colder than the air ahead of the warm front. As the cold air catches up it undercuts the warmer air. This means that the occlusion usually behaves more like a cold front than a warm front.

In practical terms, if we are waiting for a frontal system to pass through, an occlusion is likely to delay the arrival of the warm front, and more persistent rain is likely. We lose the drizzly conditions of the warm sector as the cold front combines with the warm.

The occluded front can become 'bent back' around the centre of the low, usually when the low is slow moving (*see page 27*). Where this happens the wind is likely to be light with big changes in direction, while the rain will be particularly heavy causing severely restricted visibility. The slow moving front combined with the light wind and heavy rain does not make appealing sailing conditions.

What to expect with an occluded front

Ahead of the occluded front will be similar to being ahead of a warm front, with thickening and lowering cloud.

The barometer will fall slowly.

The rain will become heavier as the front arrives.

The wind will slowly increase and may back a few degrees ahead of the occlusion before veering on the passage of the front and becoming blustery.

As the front passes through, the rain will clear and the clouds give way to sunshine and showers.

The barometer will start to rise.

When a depression passes to the south

If we are north of the depression's track we will miss the changes in weather associated with the passing of the fronts. How far to the south the low passes will determine how much cloud and rain we are likely to see; the further away from the centre, the less the cloud and rain. The wind, however, can be strong and will gradually back as the low passes. This can cause some problems if the low is tracking further south than usual, as normally sheltered anchorages may become exposed in easterly winds. The wind will continue to gradually back, as the low passes to the south; How quickly it backs will depend on the size of the low, the speed at which it is travelling, and our distance from it.

When a depression passes overhead

If we are on track for the low to pass overhead, conditions can be quite difficult. There will be thick layers of cloud and heavy rain; the wind will remain fixed in the southerly sector as the low approaches. The barometer will continue to fall until the wind lightens and becomes almost calm; this is when the centre is overhead. When the wind returns, it is from the opposite direction and will most likely build quickly. The barometer will indicate when the centre is through with a quick rise. One of the worst aspects of a low passing close overhead is likely to be a confused and uncomfortable sea as the wind changes direction. In a forecast, the wind may well be described as 'cyclonic'.

It is unusual to have the low pass directly overhead although it has happened to me a few times; the most dramatic occasion was when racing in the Southern Ocean; the wind went from 50 knots to nothing, then back to 50 knots from the opposite direction, all in the space of a four-hour watch. The sea could only be described as chaotic and rather dangerous!

Families of depressions

A depression that has developed along the polar front is unlikely to arrive on its own: it is more likely to be one of a family of depressions. This is the unsettled weather of the winter months that all too often edges its way into the summer as one depression follows another to our shores.

The first sign of a new low

On weather maps, the first sign of a new low being born is when the cold front runs parallel to the isobars. The whole system slows down and along the front the symbols change from those of a cold front to those of a warm front. The isobars separate, and on successive charts closed circulation is seen.

Between each low we often get a ridge of high pressure with light winds and some clearing of the sky. However, this can be short-lived as the next depression quickly arrives. As a rule - if, after a cold front, the showers you expect do not materialise, another depression could be on its way.

As the low occludes and declines, the cold front can often be seen to extend way out into the Atlantic. The conditions in the slow moving part of the front are very similar to the conditions that caused the first depression to form, so a disturbance along this front will cause another low to be born. This is the start of a family of depressions, each one starting life on the old cold front of the previous depression.

There are no hard and fast rules but a typical family is made up of four or five lows. Each one develops successively to the southwest and follows a track to the south of the original path of the low. Eventually the sequence is broken and the next low develops way to the north and a new generation is started. (also see also Chapter 5 and the life cycle of a depression).

Secondary Lows

A secondary low is a depression embedded in the flow around the main or primary low. As the primary low weakens, the secondary can quickly deepen and become the dominant feature, swallowing up the original low or 'dumb-belling' around it.

A secondary low may develop at any of several places, but the most common is after the depression has occluded, at the 'triple point' where the cold front, warm front, and occlusion meet (Fig 15).

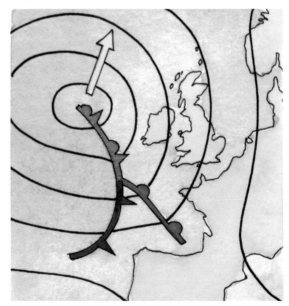

Fig 15 Development of a secondary low

Although this is not uncommon, by no means do all triple points create a secondary low. If one does occur we can see a rapid and dramatic fall in pressure over a short time, often to below that of the primary low.

On a weather chart, the first sign of a secondary low forming is the widening of the isobars before an area of closed circulation forms around the triple point (*Fig 16*).

Although the area between the secondary low and its parent usually has widely-spaced isobars and correspondingly light winds, the side of the secondary that is furthest from the parent often has much stronger winds. What makes secondaries dangerous is that they - and their associated strong winds - often develop in as little as 12-24 hours, and only after the occlusion and weakening of the parent low has lulled the unwary into a false sense of security.

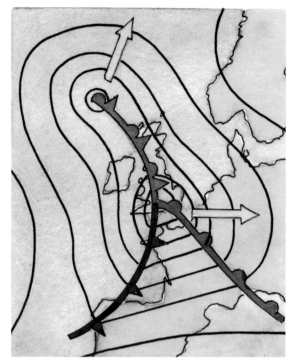

Fig 16 A secondary low developing at the triple point.

Secondary low

On a weather chart the first sign of a secondary low forming is the widening of the isobars before an area of closed circulation forms around the triple point.

As a rule of thumb the secondary low has to be greater than 600 miles away from the parent low for it to really deepen and become the dominant feature.

Anticyclones 7

The opposite of low pressure is high pressure. That may sound like a statement of the obvious, but it's important to appreciate that in meteorology, the words 'high' and 'low' are relative to the surrounding pressures, and that it is impossible to say that a pressure over XXX is 'high' or that one below YYY is 'low'.

As one might expect from the idea of a high being the opposite of a low, the circulation of air around a high is clockwise in the Northern Hemisphere (and anticlockwise in the Southern Hemisphere). This is the opposite of the cyclonic circulation found around a low, and gives us the term 'anticyclone' as an alternative, more technical-sounding word for 'high'.

Following the same theme, whereas lows are characterised by surface air flowing inwards and rising, to generate clouds, the air flow in a high is downwards and outwards, usually resulting in clear skies.

Semi-permanent highs

The global circulation (see page 13) should, in theory, produce a band of high pressure in each hemisphere, about 30-40 degrees north and south of the equator. In reality, the presence of continental land-masses breaks up these bands, leaving several distinct areas of high pressure. In the North Atlantic, for instance, there is the Azores High (or Bermuda High), while its counterpart in the south is called the South Atlantic or Saint Helena High.

They aren't permanent features: they generally move north in the northern summer and south in the northern winter, and build and decline. The position, pressure, and extent of the Azores High, in particular, has an important effect on the path of depressions moving across the Atlantic: in summer, it can become what is known as a 'blocking high', diverting depressions away from the British Isles and northern Europe to give us long periods of settled weather.

Transient Highs

More transient highs or ridges appear between the low pressure areas within a family of depressions (*Fig 17*). Although smaller and shorter-lived than semi-permanent highs, they have similar characteristics, and provide a respite from the wet and windy weather associated with the lows.

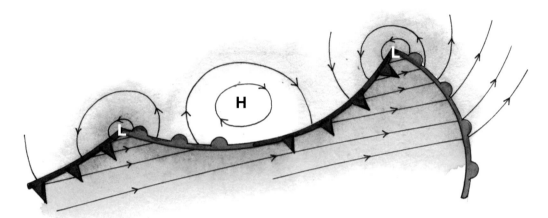

Fig 17   A transient anticyclone between two low pressure systems.

Sailing

Offshore, the wind in a high pressure area is likely to be steady. Its strength is related to the spacing between isobars, but as high pressure areas generally have widely-spaced isobars, the wind is most likely to be light, particularly near the centre. Towards the edges, though, particularly in the 'squeeze zone' between a high and a low, the isobars may be tightly-packed, with a strong pressure gradient producing strong winds.

Nearer the coast, a high is most likely to produce light winds, falling calm overnight and with the possibility of coastal fog in the early hours of the morning. Small cumulus clouds or sheets of stratocumulus may form, but the usual picture - particularly in summer - is for cloud-free days and nights, sometimes with poor visibility.

Over land, high pressure in winter is responsible for our coldest temperatures, and may give rise to a phenomenon known as an 'inversion' in which temperature increases with height instead of reducing. This traps fog, cloud, or polluted air close to the surface, producing poor visibility, anticyclonic gloom, or smog

Predicting the wind **8**

There is a definite relationship between expected wind and the spacing of isobars on a weather map, so by measuring the isobar spacing and making a few adjustments, it is possible to make a good estimate of the expected wind strength.

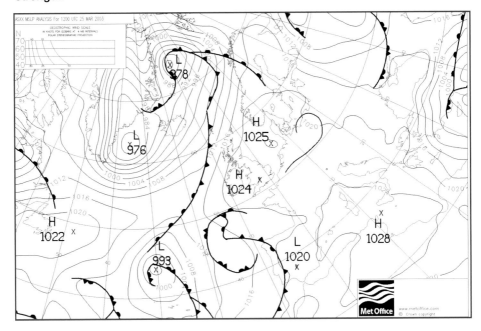

Weather maps, or synoptic charts, are generally issued in series. Each series starts with an 'analysis chart', showing the actual situation as it was recorded at various weather stations at a particular time - usually either mid-day or midnight. The series continues with charts showing the expected situation at regular intervals - usually either twelve or 24 hours - over the next five or six days.

Some sources publish analysis charts at different times; some produce forecast charts at longer or shorter intervals; and some cover longer or shorter periods. They all allow us to see how fronts and pressure systems are expected to move, and thereby predict what the weather is likely to be. It's important to appreciate, though, that the production of forecast charts is still as much an art as a science, so their accuracy almost inevitably reduces with time.

- Compare the central pressures of weather systems on successive charts to see whether they are rising or falling.

- A low generally moves parallel to the isobars in its warm sectors, but turns towards the pole once its fronts start to occlude.

- Cold fronts move at about the speed of the geostrophic wind (see page 46) measured at the front.

- Warm fronts move at about $2/3$ of the speed of the geostrophic wind.

Measuring wind from synoptic charts

Some synoptic charts have a scale called a Geostrophic Wind Scale on them. This gives an instant indication of the wind that would be blowing if all the isobars were straight and parallel, and represents the balance between the pressure gradient and the effect of the spinning earth (Coriolis Effect).

How to measure expected wind speed from the weather map

Measure the distance between two isobars with dividers (*Fig 18*); try to use a place where the isobar are straight and parallel, and measure at right angles to the isobar.

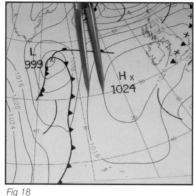

Fig 18   Fig 19

On the Geostrophic Scale (*Fig 19*), using the line corresponding to your latitude, put one point of the dividers on the left-hand edge of the scale and read off the Geostropic Wind Speed from the scale along the top or bottom edge. This can be tricky, as the scale is not linear but any mistakes made here are likely to be small compared with the consequences of using the wrong latitude; isobar spacing that gives a reading of 25 knots at 70 degrees latitude is the same as 40 knots at 40 degrees of latitude.

If the weather map is displayed on a computer screen, a piece of acetate with a scale marked on it is a good way of replacing the dividers (and is a lot kinder to the screen!).

Having measured the geostrophic wind speed, we have to modify it to get the gradient wind. The difference between the two is the result of the air curving around weather systems. Around low pressure the air slows. The tighter the curve, the more the wind slows. Conversely, around high pressure the wind speed increases - the tighter the curvature the greater the increase.

As the amount that the wind slows or increases depends on the tightness of the curvature, no adjustment is necessary if we can find a place where the isobars are straight. Very tight curvature will decrease the wind around a depression by 25% and increase it around a high by a similar amount. In extreme cases, the gradient wind around a high can be double the geostrophic wind.

Adjustments to be made

Geostrophic ➔ gradient ➔ surface

As a rule of thumb, but not as accurate as working it through:-

The surface wind speed is about $2/3$ of the geostrophic wind speed in low pressure systems and backed by about 15 degrees.

(This is for over the sea; over the land the surface wind speed is about half the geostrophic).

Measuring the wind speed from the weather map

1. Using dividers measure the distance between isobars.
2. Transfer to the geostrophic wind scale.
3. Make sure you are using the correct latitude.
4. Were the isobars straight and parallel where measured?
5. If not, adjust for curvature (- for lows, + for highs)
6. We now have gradient wind.
7. Adjust for surface friction.
8. Make allowances for gusts.

As no system is ever nicely round, it is difficult to get a very accurate result. It pays to be on the conservative side as knowing the worst the wind is likely to throw at you is safer than overcorrecting the other way. As a rule, unless the isobars are very tightly curved, restrict any adjustment to 10% for low pressure, 20% for high.

This is now the gradient wind and is the wind found above the boundary layer at around 600 metres (a height unaffected by the surface).

We now have to modify this forecast to allow for the effects of surface friction. This will vary depending on the stability of the atmosphere; here a judgment will need to be made on the air mass where you have taken the measurement as well as on the surface concerned.

The table below shows the multiplier to be used to convert the gradient wind speed to the surface wind speed, and the amount by which the direction of the surface wind should be shifted.

We now have an expected surface wind, in knots, for the time marked on the weather map.

Table for estimating surface wind from the gradient wind

Conditions	Over sea		Over land	
	Speed multiplier	Degrees Backed	Speed multiplier	Degrees Backed
Cold clear night			0.25	40
Stable	0.75	20	0.33	35
Unstable	0.90	10	0.60	20
Average	0.80	15	0.50	30

Stable - means the warm sector of a depression or a high pressure area.
Unstable - means the cold sectors of a depression.

We have almost finished but not quite. On an average day, we can expect gusts of up to $1/3$ as high again as the given wind speed. These can be stronger in unstable conditions around big showers and squalls, and less with stable conditions.

It is unlikely that our position is exactly in between isobars, and even if it is, the systems will move. We must look at the expected winds around, how the gradient changes, and whether we will we find stronger or lighter winds as we proceed.

Verifying the forecast

Weather maps cover a large area and it is a good idea to verify the analysis as best you can. On board, this is by checking the barometer, making sure it is accurately adjusted, and by noting the wind direction and strength. Our own observations of cloud structures will indicate where within a system we might be, and the likely air mass.

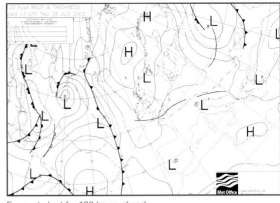

Forecast chart for 120 hours ahead

If internet access is available, you will be able to find barometer readings and weather reports from some coastal stations and airports, buoy reports, and satellite images (Chapter 14 looks at this in more detail).

Verifying the forecast. Do conditions match those expected?

Without a geostropic scale

Not all weather maps have geostrophic wind scales.

Weather fax charts, and some charts downloaded from the internet, do not necessarily have geostrophic wind scales on them. There are a number of other ways to estimate the wind from them.

The easiest is to measure the pressure gradient over a 300M distance and use the multiplier from the table (*Fig 20*). 300M is a convenient distance as it is 5 degrees of latitude, and all charts have latitude marked at either 5 or 10 degree spacing.

An example

By setting the dividers to 300M we measure the pressure change over this distance. Let us assume for our example that there is a 10mb pressure change over the 300M covering the area that we are sailing.

If this is at 55 degrees (e.g. N of the NW tip of Ireland) by using the multiplier from the table we multiply the 10mb by 2.4 giving a geostrophic wind speed of 24 knots.

Latitude (degrees)	Multiplier
60	2.3
55	2.4
50	2.6
45	2.8
40	3.1
35	3.4
30	3.9

Table for estimating geostrophic wind speed using the number of millibars over a 300M (5 degrees latitude) distance.

Fig 20

No two depressions will ever be the same. They are not neat symmetrical systems that follow the rules but are large-scale systems that cover many thousands of square miles and extend to 30,000 feet.

Varying amounts of moisture will be in the atmosphere, far more than is ever imagined. For the mathematically minded, take the size of a country, or just the county (in square metres) you live in. Add a rainfall of say 10 millimetres over the whole area, and, if your calculator can handle the zeros, an amazing volume and weight of water will have fallen.

Area of UK = 244082km^2 = 244082 million m^2 x 10mm depth = 2441 million m^3 of water.

1m^2 of water weighs 1 tonne.

The weather chart shown below shows lows and fronts that are not as neat as the diagrams in this book. Their structures will, however, be similar, and the conditions on the fronts will be as described. The wind may not veer as much as decribed in a text book and the cloud structure may look jumbled up and difficult to identify, but the basics are the same and the more we understand, the safer and more enjoyable our sailing will be.

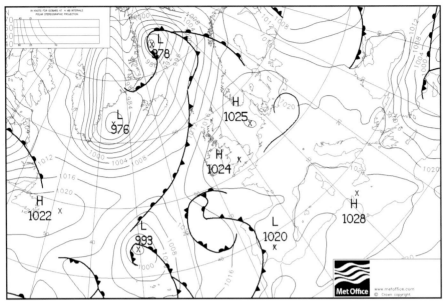

A deep low in the Atlantic with typical frontal systems. The high pressure over the UK acts as a blocking high pushing the lows north of the UK.

What forecasts mean

It is easy to get disillusioned by weather forecasts and claim that they never get it right. Sometimes this may be the case but one area of disagreement between the forecast and the sailor is about what the forecast is trying to tell us and what we are measuring and seeing.

Wind speed and direction in a forecast refer to an average wind speed over a five or ten minute period, measured at a height of 10 metres. A calibrated anemometer on a mast higher than 10 metres will read a bit more; and on a shorter mast, a bit less less. On an average day, if there were such a thing, we can expect gusts to be a third as much again as forecast, and on an unstable day, with towering convective clouds, they may be higher still.

Forecasts invariably refer to the true wind, yet while we are sailing along our instruments probably read apparent wind. Add a little tide and the instruments may very well show two forces difference on the Beaufort scale, between sailing upwind and running downwind. If you're estimating wind speed without instruments there is a tendency to overestimate light winds, and underestimate strong.

When sailing near the coast, mechanical and thermal effects of the land can have a significant impact on the wind strength. These effects are often very localised and are therefore unlikely to be included in any forecast. (See land effects Chapter 10).

It is an interesting exercise to check buoy or coastal station reports after sailing (they are available on the internet), and compare them to the winds that you have just experienced.

The Beaufort Scale

1 Light airs 1 - 3 knots.
Ripples.
Sail - drifting conditions.
Power - fast planing conditions.

2 Light breeze 4 - 6 knots.
Small wavelets.
Sail - full mainsail and large genoa.
Power - fast planing conditions.

3 Gentle breeze 7 - 10 knots.
Occasional crests.
Sail - full sail.
Power - fast planing conditions.

4 Moderate 11 - 16 knots.
Frequent white horses.
Sail - reduce headsail size.
Power - may have to slow down if wind
against tide.

5 Fresh breeze 17 - 21 knots.
Moderate waves, many white crests.
Sail - reef mainsail.
Power - reduce speed to prevent slamming
when going upwind.

6 Strong breeze 22 - 27 knots.
Large waves, white foam crests.
Sail - reef main and reduce headsail.
Power - displacement speed.

7 Near gale 28 - 33 knots.
Sea heaps up, spray, breaking waves,
foam blows in streaks.
Sail - deep reefed main, small jib.
Power - displacement speed.

8 Gale 34 - 40 knots.
Moderately high waves, breaking crests.
Sail - deep reefed main, storm jib.
Power - displacement speed, stem waves.

9 Severe gale 41 - 47 knots.
High waves, spray affects visibility.
Sail - trysail and storm jib.
Power - displacement speed, stem waves.

10 Storm 48 - 55 knots.
Very high waves, long breaking crests.
Survival conditions.

11 Violent storm 56 - 63 knots.
Exceptionally high seas with continuously
breaking waves seriously affecting visibility.
Survival tactics.

12 Hurricane 64 knots and above.
Exceptionally high seas with continuously
breaking waves seriously affecting visibility.
Survival tactics.

The Beaufort Scale was developed by Admiral Francis Beaufort and was first published in 1808, that was long before anemometers were available, in an effort to standardise weather reports, Beaufort's scale linked wind strength to the sea state and the amount of sail that could be carried by the ships of his time. This very human and easily understood scale is still the most widely-used way of expressing wind strength in marine weather forecasts.

Nowadays the Beaufort Scale still links wind speed with the sea state that can be expected in open waters away from the influence of land or tidal flow.

The forecast

The wind direction given in forecasts refers to where the wind is coming from, i.e. a northerly wind is blowing from north to south. Only the main eight points of the compass are generally used: dividing it up much more implies an accuracy that is unlikely to be met over the area of the whole forecast. Reports are sometimes given in degrees (true) or using a 16 point compass, making it a useful exercise to look at the chart and use the compass rose on it to see the direction.

A lot of information is given in a few words in the shipping forecast and precise definitions are used to remove ambiguity. Forecasts start with any gale warnings - issued if the mean wind speed is expected to reach Force 8 (or if gusts are expected to reach Force 9). Severe gale Force 9 implies a mean speed of a Force 9 or gusts of Force 10, and so on through to hurricane force. Radio 4 will also issue gale warnings outside the forecast times, when received, and then repeated on the hour, ahead of the news.

Timing of gale warnings is very important although accuracy in timing is a part of forecasting that is particularly difficult. Our position within the sea area will make quite a difference depending from which direction the depression is coming.

Timing of gale warnings	
Imminent	Within 6 hours
Soon	6 - 12 hours
Later	> 12 hours

In coastal waters forecasts, strong wind warnings may also be issued when the wind is expected to reach a Force 6, which is usually more than enough for good sailing.

The general synopsis follows any gale warnings and gives a general picture, describing the weather chart in a few words. The main features of low or high pressure will be included, as well as their movement. Again, terms with precise meanings are used to describe the likely movements.

From the time of issue for gale warnings

Synoptic system movements	
Slowly	< 15 knots
Steadily	15 - 25 knots
Rather quickly	25 - 35 knots
Rapidly	35 - 45 knots
Very rapidly	> 45 knots

There is a lot of useful information given in the general synopsis and with practice we can use it to help explain the weather we are about to get.

Sea area forecasts

The next section is made up of the sea area forecasts. There is one big rule here and that is know which area you are in! It may seem obvious but it is important - and so is the idea that the weather is a moving entity, which cannot be put into neat boxes. Take note of adjoining sea areas to gain a better picture.

With the wind strengths and directions also comes a description of the weather; within this there will be an indication of rain. As has already been explained in the chapter on clouds, showers only occur from convective clouds: on the whole they are likely to be heavier than other rain, but last for a shorter time.

The type of rain helps to locate where we are within a depression. Rain from layer clouds implies being ahead of the warm front, but mist and drizzle are indicative of the warm sector. Showers and rain will also adversely affect visibility.

Visibility, however, has its own place in the forecast, described as below.

Visibility	Distance
good	> 5M
moderate	2 - 5M
poor	1,000m - 2M
fog	< 1,000m

Sea areas

Pressure changes	
Description	**Change in 3 hours**
steady	< 0.1mb
slowly	0.1-1.5mb
rising/falling	1.6-3.5mb
quickly	3.6-6.0mb
very rapidly	> 6.0mb
now rising	the trend has reversed
now falling	the trend has reversed

After the early morning shipping forecast we then get reports from coastal stations. These are actual conditions at the time of the report and most are self-explanatory. They include the pressure reading and a useful indication of how it has changed over the past 3 hours.

All the above describes what could best be called a generic maritime forecast - as used in the BBC Shipping Forecast but it is similar to forecasts transmitted throughout the world by radio or text. Through the World Meteorological Organisation many standards have been set and with just a few exceptions, units and words have been standardised. One important difference is that some Scandinavian and European countries report wind speeds in metres per second, instead of knots. Double the metric figure to give an approximate conversion to knots.

As discussed later in this book, there are many local influences that change the wind we get, and it would need an unrealistically detailed forecast to try to cover all eventualities. Forecasts are generally for a large area – sea area Shannon, for instance, measures about 210 x 200 nautical miles but is covered in just a couple of sentences for a 24 hour period.

Taking a flight across the Irish Sea or the Channel is a revelation; looking down on the variety of clouds shows how different the weather can be in places just a few miles apart. Satellite pictures are excellent for this and are now familiar from TV weather broadcasts.

If you are coastal sailing, the inshore waters forecast should also be taken into account as it covers from the coast to 12 miles offshore and is repeated by the Coastguard on VHF.

Coastal waters forecast

The coastal waters forecast splits the coast up into 16 sections; an example for one area as published on the internet follows:

Colwyn Bay to the Mull of Galloway including the Isle of Man 24 hour forecast:

Wind: east or southeast 3 or 4, occasionally 5, increasing 4 or 5, and occasionally 6 perhaps 7 in north later.

Weather: fair.

Visibility: good.

Sea State: slight building moderate in north.

Outlook for the following 24 hours:

Wind: east or southeast 4 or 5, becoming mainly southeast and occasionally 6 perhaps 7 in north.

Weather: fair, some haze.

Visibility: good, perhaps moderate.

Sea State: slight in east, moderate in north, perhaps rough later.

Definitions are the same as before with a few extras. Under the heading of 'weather' for instance, 'fair' means that there is no significant event happening. It may be cloudy or sunny but there should not be any rain or showers.

Sea state definitions are included,despite the fact that conditions may vary considerably because of the orientation of the coast, headlands, tides and water depth.

The coastal waters forecast still covers a large area and in our example we have the high land of the Isle of Man in an area stretching from Wales to Scotland.

Wave height	
Sea state	**metres**
calm	< 0.1m
smooth	0.1 - 0.5m
slight	0.5 - 1.25m
moderate	1.25 - 2.5m
rough	2.5 - 4m
very rough	4 - 6m
high	6 - 9m
very high	9 - 14m
phenomenal	over 14m

It is all but impossible to improve on the professional forecaster and the huge computing power of the Met services when looking at the synoptic situation. Our knowledge of systems, and the theory covered in the first part of this book, is to help put detail into a forecast that covers a large area. It will help us to understand if things are not going to plan, and most importantly, using our understanding of the forecasts, we will be able to modify them for where we are sailing.

Effects of land on wind

Having listened to the forecast, downloaded synoptic charts from the Internet, or received them by fax, it is now time to add our own input into the picture that makes up the weather jigsaw.

Almost by definition, our starting and finishing points are sheltered. The passage between them often follows the coast, so unless we are in the middle of a sea area, well away from the land, we must always consider how the proximity of land is likely to modify the forecast. Just how far away from the land we need to be for the wind and weather to be unaffected is a little like 'how long is a piece of string?'.

Ten or twelve miles is a good guide, but it varies depending on whether the wind is blowing from the shore, onshore, or parallel to the coast. A good sea breeze in the spring or early summer may extend even further. Mountains and high land in the stream of wind may influence the wind for many miles, as anyone caught in the wind shadow of the Canary Islands will know. The Mistral at times can affect most of the Western Mediterranean. In rivers and landlocked venues we need to look at the land forecast and modify it for over the water, whilst from the coast we can use the sea forecast and modify it for the influence of the land.

Air behaves like water and looks for a path of least resistance. Watch a stream; the faster the current and the greater the number of rocks and obstructions, the greater the disturbance to the general flow. The wind is like this on a much greater scale, and thinking of it in these terms can help in understanding why the wind behaves as it does around land.

How the land modifies the wind can be split conveniently into two processes - mechanical and thermal effects.

Friction

We have already discussed the fact that the wind has a direct relationship with the pressure gradient – represented on weather maps by spacing between the isobars. This, however, is the gradient wind at about 600m (2,000ft) where it is undisturbed by the surface. In Chapter 8, we looked at ways of estimating the gradient wind from weather maps.

The Coriolis effect tends to make the wind at higher levels (above about 600m) blow parallel to the isobars. At ground or sea level, however, the moving air is slowed down by friction, and the Coriolis effect is reduced so the wind is diverted more towards the centre of low pressure.

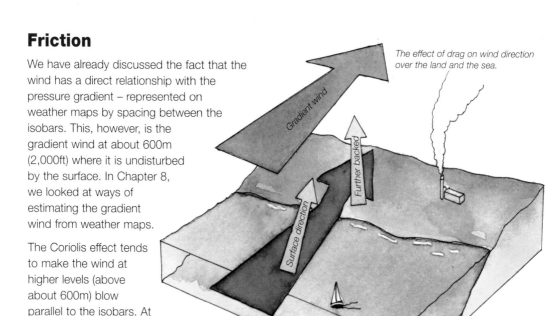

The effect of drag on wind direction over the land and the sea.

Gradient wind

Further backed

Surface direction

Fig 20

How much the wind is slowed and deflected depends on the roughness of the surface and on the buoyancy of the air. Over land, there is a big difference between smooth grassy fields and the roughness of forests or cities. Forests or cities reduce the wind speed more than open fields, and increase the extent to which it is backed. By comparison, even a 'rough' sea is effectively smooth, so the wind is slowed down and diverted to a much lesser extent.

As a rule of thumb, the surface wind over the sea will be about 80-90% of the gradient wind, and will be angled inwards from the isobars by about 10-15 degrees. Over land, it will be about 50% of the gradient wind, and will be angled inwards by about 30 degrees. The change of wind speed and direction is not instantaneous: an offshore wind doesn't immediately change direction and increase in speed as it crosses the coast. This has important implications for coastal sailors.

> As a rule of thumb, in average conditions the surface wind over the sea will be about 80-90% of the gradient wind, whilst over land it will be closer to 50% of the gradient wind.

The difference between the surface wind and the gradient wind can often be seen from the movement of low clouds. If you stand facing the surface wind, the clouds (in the Northern Hemisphere) will usually be seen approaching from your right, because they are being blown along by the gradient wind.

> When the wind blows off the land it will increase and veer over the sea.
>
> Conversely, as the wind blows from the sea over the land it slows and backs (see *Fig 23* on page 68).

Stability and instability

Stability and instability refer to the relationship between the air near the surface and the air above it. Temperature is a particulaly important factor.

It is important to remember that the air is heated or cooled from below.

The air is not affected by light from the sun shining through it, but is changed by the temperature of the surface below. This therefore heats or cools the air from the lowest levels upwards, thus changing the buoyancy of the air, causing it either to rise or sink.

If the air is being heated from below, it will rise - along with the water held in it - giving a buoyant or unstable atmosphere. If, however, the air at surface level is being cooled it will stay where it is and the air will to be stable.

Stable conditions	Unstable conditions
Steady surface wind. Layer clouds – stratus or no clouds at all. Poor visibility and possible fog. Typically - warm sector or high pressure.	Gusty conditions. Cumulus clouds and any rain will be showers. Good visibility outside of showers. Typically - after a cold front, thunderstorms.

Over land the surface temperature has a strong daily cycle as the land heats during the day and cools at night. Over the sea, however, there is little change in the temperature throughout a 24-hour period.

In the absence of a change of air mass from a synoptic scale system, or strong winds, the temperature and wind changes throughout the day are shown in *Fig 21*.

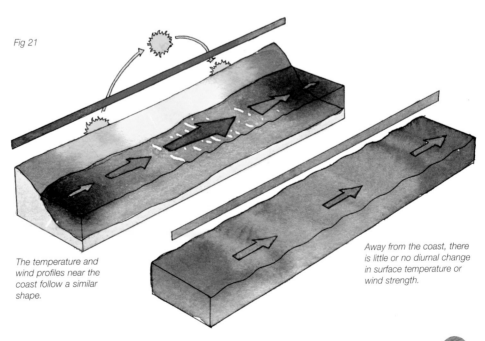

Fig 21

The temperature and wind profiles near the coast follow a similar shape.

Away from the coast, there is little or no diurnal change in surface temperature or wind strength.

Thermal effects

Fast-moving deep depressions and large-scale synoptic conditions will override local thermal effects but, on the majority of summer days, there will be some thermal influence.

We have all seen days when it is flat calm early in the morning, but by mid-afternoon we are reefed down, only for the wind to drop again at night (see *Fig 22*). This is the typical wind pattern throughout the day near coasts, particularly in the summer, when there are no large weather systems around bringing cloud, strong wind, or changes in air mass.

Over the land, the temperature rises to a peak in the early afternoon and falls to a minimum at around dawn.

This diurnal change in temperature means that the land is warmer than the air during the day but colder at night. This has the effect of changing the stability of the air from unstable during the day, to stable at night.

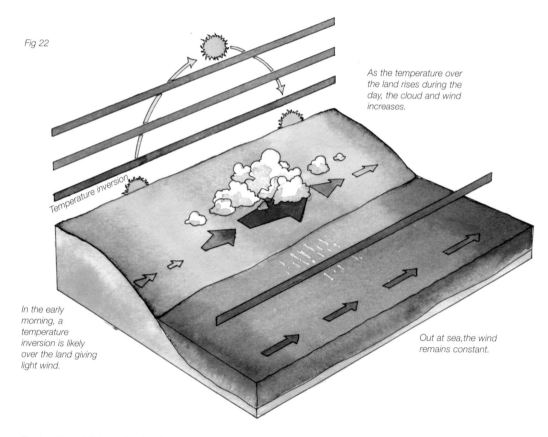

Fig 22

As the temperature over the land rises during the day, the cloud and wind increases.

Temperature inversion

In the early morning, a temperature inversion is likely over the land giving light wind.

Out at sea, the wind remains constant.

During the night, as the land cools, it in turn cools the bottom layers of the atmosphere, making it become more stable. We have already seen that the more stable the air is, the greater the effect of drag on the wind. This is the start of a feedback mechanism; the land cools the air, it becomes more stable and is slowed by friction, so the cooling earth has a greater effect on the air, cooling and slowing it still further. This can stop the wind completely, often giving calm conditions over land on clear cold nights. Gravity helps drain this cold air down rivers and out over coastal waters creating a band of light wind around the coasts.

This continues until dawn, which is the coldest part of the night. It also creates an inversion in the temperature at the bottom of the atmosphere, with the air getting warmer with height rather than colder. Above the inversion and the boundary layer, the gradient wind is still blowing – indeed there may be a stronger band created above the boundary layer (known as a nocturnal jet).

Calm conditions often found at night

Obviously the clearer and colder the night, the greater the chance of the wind dropping. The effect spreads to coastal waters giving quiet nights at anchor when cruising in the summer.

Once the sun gets to work heating the land, the inversion begins to break down. It does, however, need the vertical movement of the warming air to break the inversion and get the flow of air mixing again before we will feel any surface wind. A sign of the inversion breaking down is the development of cumulus clouds over the land.

As the day progresses and the ground warms, the boundary layer begins to heat. This reduces the stability giving more vertical movement to the air. This mixing increases the surface wind as the land heats and some of the gradient wind finds its way to the surface.

The sea breeze

In some form or another, the sea breeze can be found during Spring and Summer throughout the world. It can be a regular feature like the Fremantle Doctor of Western Australia or the afternoon breeze that makes for a perfect Cowes Week.

Any onshore wind is usually termed a 'sea breeze' and because, like a sea breeze, most days are windier in the afternoon, it is not always possible to determine what is a sea breeze and what is either a change in the synoptic situation or an enhancement of the gradient wind. Understanding the differences between an onshore wind and a true sea breeze however, will help to plan the day's sail, and will greatly improve your chances on the race course.

Conditions necessary for a sea breeze to develop

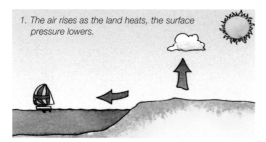

1. The air rises as the land heats, the surface pressure lowers.

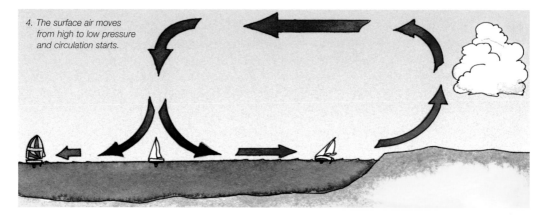

2. Helped by an offshore wind, the excess air spreads out over the sea.

1. The land must heat up to a greater temperature than the sea water – hence the best sea breezes are in Spring when the water temperature is at its coolest and in Summer when the land heats up the most. Local land radio broadcasts give expected temperatures.

2. There needs to be a light gradient wind that has an offshore component to it. 15 knots is generally considered the maximum gradient wind for the sea breeze to overcome. The greater the temperature contrast between land and sea, the higher the offshore wind the sea breeze can overcome.

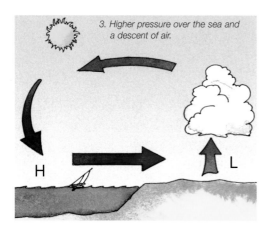

3. Higher pressure over the sea and a descent of air.

H

L

3. There needs to be convection to a moderate level. A difficult one to define, convection is needed to get the circulation going and can be seen by cumulus clouds forming. Too much convection, producing towering cumulus clouds, showers and possibly thunderstorms, can kill the sea breeze. Not enough, which can be the case during high pressure, and no matter how hot the land becomes, a sea breeze will struggle to develop.

4. The surface air moves from high to low pressure and circulation starts.

4. The sea breeze develops as the land heats, which in turn causes the pressure to fall. The excess rising air drifts out over the sea – a light offshore gradient wind helps in this, increasing the pressure differential. Slightly higher pressure over the sea and lower over the land starts a circulation, as the surface air moves from high to low pressure.

A good sea breeze day with clouds over the land

The sea breeze (in theory) will start near the shore and blow at right angles to it. As the day progresses it will keep extending offshore, strengthen, and veer (or back in the Southern Hemisphere), eventually blowing at about 20 degrees to the coast. It can reach some 20 miles out to sea and penetrate considerably more inland, but as no shore line is straight, no sea breeze adheres strictly to the rules.

The sea breeze will be accelerated around headlands and funnelled by the land (as explained on page 70). This can give a sea breeze that reaches 20 or even 25 knots in places, but in general, rather less. The land behind the coast is important, as steep barren mountains will heat more quickly than vegetated fields. The sea breeze, although possible to predict, varies greatly depending on locality.

Wind through a typical sea breeze day

Light offshore wind in the early morning.

A period of calm, mid-morning, starting close to the beach.

An onshore wind develops near the coast.

The band of onshore wind rolls seawards but remains strongest near the land.

The wind veers as the afternoon progresses (backs in S Hemisphere).

Maximum wind speed mid to late afternoon.

It is the surface pressure differential between the land and sea, created by the heating effect of the land, that starts the sea breeze off. We have also seen that an offshore component to the wind helps to get the circulation going.

If we have high pressure over the land and lower pressure over the sea, determined by using Buys Ballot's law (*see page 28*) or by studying the charts, the chance of a sea breeze diminishes. This is because the pressure differential has to be reversed before it is possible to get the circulation going. This will normally cause the wind to drop as the pressure equalises, with only a late, light sea breeze developing - if one develops at all.

The clouds through a sea breeze

As we need convection for sea breezes to develop, we can expect cumulus clouds to appear. The earlier we start to see the cumulus develop, the earlier a sea breeze is likely to start. A clear patch of sky over the water with the clouds lining up on the land side of the coastline is often the first sign of the sea breeze.

As the sea breeze develops, the cloud front moves inland and the clear area extends seawards. This is one reason why seaside resorts tend to be the sunniest places, as sea breezes move the clouds away, whilst only a few miles inland the clouds develop. We are even luckier sailing in coastal waters, looking at clouds over the land whilst sailing in clear skies.

Not quite a sea breeze

Onshore gradient winds will often strengthen during the late morning and afternoon by 5-10 knots. This is caused as the land heats and the pressure falls. This is not a sea breeze but a thermally-enhanced gradient wind. Picky perhaps, but it has important differences:

- there will not be a period of calm before the wind increases.

- the wind is unlikely to veer (in fact it may back as the pressure over the land falls).

- the cloud pattern is different.

- the clouds may not clear over the sea as there is no circulation bringing descending cooler air to clear them.

If the onshore gradient wind is from a direction that implies higher pressure over the land, then the wind is likely to drop during the day, however hot it becomes over the land.

Islands

Islands are interesting because they generate their own sea breezes. Small islands close to the coast will be overwhelmed by the sea breeze generated

Clouds building over the mainland shore.

by the larger land mass, although initially an island sea breeze may develop. The Solent can be an interesting place for sea breezes; sometimes spinnakers can be seen coming up the Western Solent on a southwesterly breeze at the same time as spinnakers are seen arriving from the east on a southeasterly. Between the two, calms are found over the Brambles Bank. Usually, the southwesterly breeze eventually dominates the whole area.

Larger islands, for example Majorca, see sea breezes form on all coasts. This makes it hard to find afternoon anchorages, as all coasts become lee shores, and as the wind blows strongest near the land, bays that might be expected to be calm are not.

With any island or peninsula, sea breezes that meet from different coasts give a convergence zone, where the only way for the air and moisture to go is upwards. This gives towering clouds and showers inland where they meet, often enhanced by hills or mountains.

Offshore night breezes

The opposite of sea breezes, these are more influenced by topography than anything else. The air cools and drains by gravity to the sea, so the strongest night breezes can be expected opposite valleys and estuaries. These winds can be very localised and, near mountains, are as regular as a sea breeze.

It can be argued that a reverse circulation to the sea breeze is created, but this will only give a very light land breeze, the main effect being to bring the cold stable air off the land and over coastal waters, resulting in mainly calm conditions.

Katabatic winds are cold mountain winds blowing down valleys, and may spill over to the coast. They are driven by gravity and in high latitudes reach gale force or stronger. (Anabatic winds are the opposite, driven by rapid heating of rocky mountains creating circulation).

If sailing at night near mountainous regions, particularly if they are snow-capped, expect strong, cold, blasts of wind. Lake sailors are familiar with both katabatic and anabatic winds.

Mechanical effects

Sailing near the coast we find bands of wind that change considerably in both strength and direction. It is a rare day indeed when the wind is steady, and this is generally only when it blows directly onto the shore.

Stability is one reason why there are gusts and lulls, and although we have looked in detail at how the stability changes over land due to the changing temperature, we must also look at the different air masses and how stability affects them.

The changes along the coast are often greater than explained before and are due to mechanical forces as the wind flows over, and around, obstructions seeking a path of least resistance. There are also effects due to coastal divergence and convergence.

Most coastlines change direction with bays and headlands, estuaries and islands, making it necessary to look at the principles of how the wind is changed by the coast line.

Some features will affect the wind for many miles downwind, whilst others last for just a few hundred yards. The effect of funnelling is an example where the wind increase lasts for many miles.

Through the Straits of Gibraltar the wind is bent and accelerated by the high mountains to the north and south. This funnels the wind, making entering or leaving the Mediterranean difficult. This has, however, put Tarifa on the windsurfing map as a good place for strong winds!

On a smaller scale, the wind is often accelerated at the entrance of a river although all around the wind is light.

With an offshore wind

When the wind blows from the land to the sea it is said to be offshore.

Friction slows and backs the wind to a greater extent over the land than the sea (see page 60). It therefore follows that as the wind leaves the land it veers and accelerates (Fig 23).

As a guide, the wind will veer about 15 degrees and increase considerably, sometimes doubling in speed as it leaves the land.

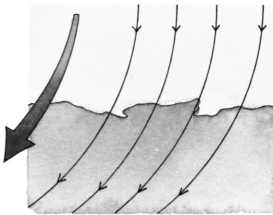

Fig 23   The wind veers and accelerates as it leaves the land

The effect of the wind increasing as we move away from the land is most noticeable in the first few hundred metres but it can continue for several miles out to sea. Few coastlines are uniform, so the wind will funnel down any valleys, creating bands of stronger, gustier wind that extend out to sea.

This makes it difficult to predict the wind at sea from the security of a sheltered marina.

The bend in the wind as it leaves the land can be used to advantage, as a steady lift can be expected on port tack, when beating towards the land, as well as a decrease in the wind. However, a river in line with the general wind flow direction is likely to funnel the wind, giving a stronger wind and a dead beat. The opposite is true when leaving the shelter of the land and an early reef is a wise precaution, as the wind strengthens considerably away from the land.

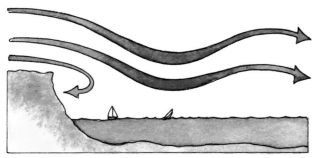

Fig 24

Wind coming off the land over cliffs can give unexpected patterns. Close to high cliffs, eddies can form in the wind flow, giving a reversal of direction (*Fig 24*). The wind beyond this is then often strongly banded as the air forms a wave, giving stronger wind as it reaches the surface. Known as lee waves, they can also be found wherever mountains lie across the flow.

With a wind blowing parallel to the shoreline

Coastal convergence and divergence

When the wind is blowing parallel to the shore, the wind over the land is backed with relation to the wind over the sea. Consequently, on a straight coast the wind over the land bends either towards the wind over the sea – known as convergence, or away from it – known as divergence.

If the land is on the starboard side of a boat running with the wind astern, the wind will be converging so, for a band of up to two or three miles from the coast, the wind will be strengthened by a few knots – perhaps as much as 25%.

If the land is on the port side of a boat running with the wind astern, the wind will be diverging, so the wind will be lighter along the coast.

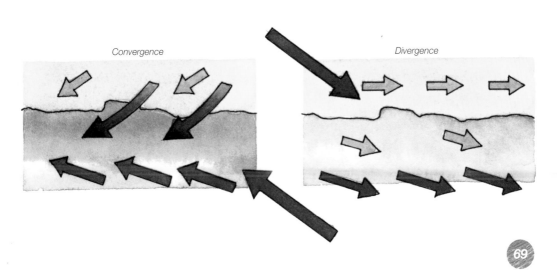

Convergence

Divergence

When sailing near the coast with the wind blowing parallel to the land

With the wind on your back, land on your left, expect less wind.
With the wind on your back, land on your right, expect stronger winds.

No coastline is straight and the wind is rarely parallel to it, so the effects are not uniform. Bands of stronger and lighter winds may extend a couple of miles out to sea but there are no hard and fast rules. The biggest help is in knowing what to expect when approaching the coast, whether the wind is likely to increase or decrease, so that sail selection can be made in good time.

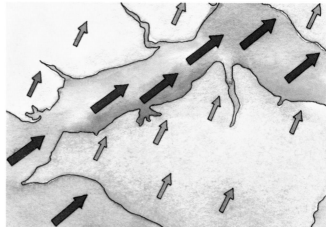

Convergence and divergence in the Solent.

Headlands

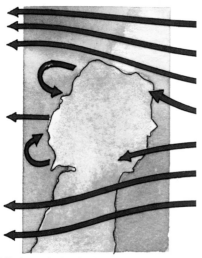

Wind around a headland.

Any headland jutting out into the flow of wind will funnel and bend the wind around it. There will be an area of acceleration where the wind flows come together (convergence) and a fanning out of the wind behind the headland (divergence). The terms are the same as when talking about coastal convergence. When headland convergence and coastal convergence work together, the wind may increase substantially. If the headland is high, lee eddies may form along with gusts and lulls.

If the headland is on a convergent coastline, then the increase of wind around the headland will be greater, as the backing wind passing over the land will add to the wind increase. No two headlands are the same and a small change in the wind direction may make a big difference, as does the actual shape and height of the land.

On a divergence coast, the lighter band of wind will be expected close in, and although there is likely to be an increase in wind around the headlands due to funnelling, the overall increase should be less than on a convergent coastline.

A side effect of convergence and divergence is in the amount of cloud. Converging airstreams create an ascent of air and will create clouds, whilst diverging airstreams create descent and less cloud.

Onshore

Where the wind blows directly onto the land

With the wind onshore, there are few effects felt out on the water as the changes in the wind occur over and around the land. However, close to cliffs or high obstacles on the coast, the wind will lift off the surface leaving an area of flukey wind close inshore (*Fig 25*).

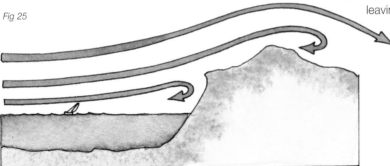

Fig 25

There can also be a change in the wind near headlands jutting out into the flow. Wind bands will be stronger either side of the headland as the flow of wind is divided by the land.

Island in the flow

The divergence and convergence argument can be taken further and applied to islands. An island in the flow will have a windy side and a calmer side depending on the wind's direction.

The height of the island will also have an affect, creating gusts and lulls. *Fig 26* shows a neat band of stronger wind, which in reality will not be as clear-cut as the diagram suggests.

The land effects, whether thermal, mechanical, or a mixture of both, help to explain why the wind that we are actually sailing in can be quite different from that which was suggested by a wide-area forecast. No forecast can take into account the local variations in the wind that can be found just a few miles, if not metres, apart. Predicting local variations is not only interesting to do, it will also add to your enjoyment of being on the water.

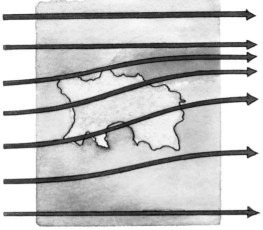

Fig 26   Convergence by island showing a windy site and a calmer side.

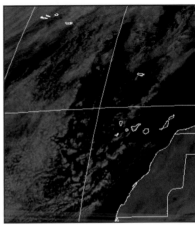

The Canary Islands are well known for funnelling, and convergence zones. The high islands disturb the flow as shown in the cloud patterns.

All at sea 11

Once we are away from the influence of the land, the winds will be more like those we might expect from the synoptic chart. This does not, however, mean that the wind will be steady in either direction or strength, and there are still likely to be bands of wind of varying strength.

Trade wind sailing

There is still the need to look at the forecasts to see the overall synoptic picture before refining it for the area we are sailing in. We need to know how the wind is likely to behave in both the short and long term:-

Short term - to decide which sails to use and the conditions to prepare for.

Long term - to position the boat to take advantage of any wind swings or to avoid, if possible, strong winds.

Weather routeing

Although it is thought to be the preserve of single-handed ocean races or merchant shipping, we all weather route to some extent whenever we leave harbour. Heading up a few degrees in case we get headed when making for an entrance, or deciding what port to head for, is all weather routeing but on a less grand scale than having a weather forecaster send directions.

Away from land, on a long crossing, there is greater flexibility in our course and it may be worth sailing extra miles to stay with favourable winds.

Planning

Whilst we are still in the planning stage ashore, synoptic charts can be freely downloaded from the internet, received by fax, or found on notice boards in marinas.

Synoptic charts are our basic tools and we can add to them in a number of ways, particularly if we have internet or email capabilities. (See Chapter 14).

The analysis of synoptic charts brings together the theory of synoptic scale systems and the likely cloud and weather to be expected. We have seen how to measure the wind speed from the chart and modify it for the drag of the surface. We should add our current observations to confirm that what should be happening and what is happening are the same.

All the time we are adding to the jigsaw to understand what is happening, in a meteorological sense, so that we can forecast the most likely changes and use them to our advantage. Knowing if the wind is going to change in direction or strength allows us to head off on the most favourable tack.

Over the open ocean, without the influence of land, there are still localised changes to be expected in the wind. These will be mainly associated with clouds or a change in water temperature.

Large changes in water temperature are linked to main ocean currents, and are well documented in pilot books and on routeing charts. Routeing charts give average monthly meteorological conditions to be found over an ocean and include major currents.

Ocean currents are important as they move cold or warm water around the globe and help to regulate climate; the water modifies the temperature and humidity of the air above it. The boundaries between warm and cold water are often places of significant meteorological events where cyclogenesis (development of depressions), fog and thunderstorms are likely.

Some of these features are discussed further in Chapter 12 (Meteorological dangers) and Chapter 16 (Around the world).

The boundary between warm and cold water will also divide air of different stability. The warm water heats the air making it more unstable, and the circulation brings more air from above the boundary layer, reducing drag and increasing the average wind speed.

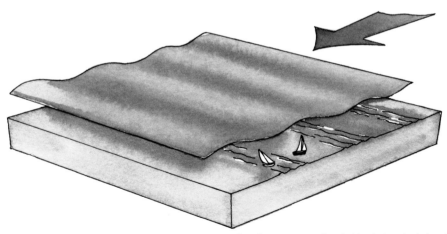

Over the open ocean the wind tends to arrive in bands

There will also be more gusts and lulls with the wind veering and backing a few degrees as each gust passes. If this sounds familiar it is because the boundary between the water temperatures acts like a coastline, with slower backed wind over the colder water, and stronger veered wind over the warm.

Wind Lanes

Over open water, particularly in the trade wind belts, fair weather cumulus clouds often form in well-defined bands, or lanes, with clear sky between each lane. Cumulus clouds are a sign of ascending air, so under each line of cloud there is likely to be a band of relatively light winds, where the surface air is rising.

The clear bands, by contrast, are a sign of descending air. The air aloft is usually moving faster than the surface wind, so when it drops to surface level it will be felt as an increase in the wind strength. When it reaches the surface, of course, it can fall no further, so it moves horizontally, to replace the air that has been drawn upwards under the clouds. Where this horizontal movement is in the same direction as the gradient wind, they combine to produce a band of wind that is stronger still.

The whole pattern of alternating bands of strong and light winds tends to move as though it were being blown along by the gradient wind. This means that if you are trying to stay in the strongest wind, it is better to sail through the lulls on port tack, in order to reach the next clear band as quickly as possible, and then to tack onto starboard in order to stay with it.

Any change in the clouds is significant and needs to be identified and fitted into the synoptic situation. Small-scale features will not be included in the synoptic charts, as they are too small to be represented but they are still significant to sailors. These small scale features include thunderstorms and squalls which can add significantly to the wind strength.

From novice crew to superyacht skipper, one of our biggest worries is getting caught out in bad weather, at some time in our sailing career we are likely to find ourselves in conditions that we would rather not be in.

Gales

Gales do not arrive without any warning at all, but sometimes conditions may deteriorate more than expected. One of the best instruments, but one that is often overlooked in our high tech world, is the barometer. There may be local conditions that accelerate the wind (see Chapter 10) but when a gale develops it will invariably be preceded by a change in pressure. This change need not be a fall: as the old saying goes, *a quick rise after low predicts a stronger blow.*

As a rule of thumb, a change of 5 or 6mb in three hours predicts a Force 6, whilst a change of 8mb foretells a Force 8. This only a guide, as a lot depends on the wind we have as our starting point. If it is already blowing a Force 6 or 7 and the barometer is dropping quickly, then a gale is imminent. Starting at a lower wind strength we have a bit more time to reach shelter. The same is true with a quick rise in pressure - any steep pressure gradient indicates strong winds.

In many depressions, the tightest isobars, and hence the strongest winds, are found on the back (northwestly side) of the low. If the low is moving in the customary way of west to east, the stronger wind soon passes.

If in the warm sector of a depression the barometer is still falling, the low is either deepening or moving closer – in either case expect stronger winds.

Secondary lows

Secondary lows can develop in the circulation of a main depression, with pressure dropping quickly as the secondary low intensifies to become deeper than the primary. This rapid deepening is sometimes described as a 'bomb' or 'explosive' cyclogenesis and normally only happens if the secondary low develops more than 600M from its parent

The strongest wind will be on the side away from the original low. The first signs on a synoptic chart are the widening of the isobars before successive charts show circulation and a deepening of central pressure. On satellite pictures, a comma-shaped cloud develops and this can be the first real sign that the low is deepening quickly.

Gusts

Gusts are increases in wind – rapid but short-lived. The opposite of gusts are the lulls which accompany them. There are a number of reasons why gusts occur; if the wind is blowing off the land they are probably of a turbulent nature caused by obstructions to the wind on the land. In an onshore wind, they are more likely to be caused by vertical movement in the atmosphere and are likely to be strongest on a day when the atmosphere is unstable. This is indicated by large cumulus clouds: the greater the vertical extent, the gustier the conditions.

Gusts caused by instability of the atmosphere tend to veer the wind when they arrive and may increase wind strength by 30 to 50%.

Boat broaching. Gusts can catch even professional sailors out!

Gusts generated by turbulence over land can increase the wind strength even more, sometimes as much as doubling the average.

Squalls

A gust becomes a squall if it lasts longer than a minute and involves a wind speed at least 16 knots higher than the mean, and exceeds 22 knots. It is obviously a significant event and potentially dangerous. The warning signs are usually rain or hail, and a big dark cloud.

The term 'squall line' was originally used to describe a cold front - it is a good description. Cold fronts, or troughs, can be seen approaching with towering cumulus clouds and

heavy rain. Expect squalls to back the wind slightly as they approach and to veer the wind once they pass.

Squalls are common in the unstable air behind cold fronts, particularly if the air is very cold. If there is precipitation falling from a big cumulus cloud, expect the worst and reef early.

Squall clouds have their own circulation similar to that for a cumulus cloud. They are much bigger, reaching to great heights and are steered by the wind blowing at a much higher level. They are similar to the thunderstorms described below.

Not all large black clouds have squalls attached and it is sometimes difficult to know which will produce a squall and which will just give a small short-lived gust. Rain is a good indication; the heavier the rainfall, the stronger the squall is likely to be.

Thunderstorms

Thunderstorms are not only dramatic but can also be dangerous. There is the danger of being struck by lightning (luckily, this only happens rarely) but there is also the risk of a strong gust of wind ahead of the storm arriving with gale force ferocity along with stinging rain and hail.

Thunderstorms are caused by instability in the atmosphere, with overheated air near the surface and much cooler air above. This shows itself as towering cumulus clouds – which may or may not have the classic 'anvil shaped' head described in many text books. As the clouds grow, a thunderstorm – or at least a heavy squall, is possible.

The gust is short-lived but potentially damaging in its violence and, as it arrives quickly, gives little time to shorten sail. It soon gives way to heavy rain and a dropping wind. The wind direction is likely to back ahead of the storm and veer in the strong gust.

Thunderstorms reach up to the tropopause, which may be some 12km above our heads. The wispy anvil top that is sometimes seen is ice crystals blowing in the jet stream at that height. Because of the vertical extent of the clouds, the movement of the storm follows the winds high up. This may be very different from the direction of the wind that we are sailing in so the track of a thunderstorm or big squall may be difficult to predict and needs close monitoring.

Luckily, thunderstorms are more common over land, although they can drift over coastal waters. The UK only has a few days a year when thunder is heard, but in hotter climates, they may be more common.

Altocumulus castellanus are a sign of thundery weather. As the name suggests they are lumpy clouds that reach high into the atmosphere and look (loosely) like castle ramparts in the air.

In the western part of the UK, thunderstorms are mostly associated with the passing of active cold fronts, but the south and east can suffer from thundery troughs coming from the continent. These are usually associated with hot weather which is why the phrase *three fine days and a thunderstorm* has been used to describe a British summer.

There are internet sites that show daily lightning strikes, known as sferics, and it is quite interesting to see where the thunderstorms commonly occur (Web sites are given on page 103)

Thunder clouds - cumulonimbus

Thunder, incidentally, is the sound produced by lightning, and they happen at precisely the same moment. The reason thunder is usually heard after lightning is because sound travels more slowly than light. As a rule of thumb; the time difference between lightning and thunder is about six seconds per nautical mile. This is a good way to monitor the progress of a storm.

Fog

There are two main types of fog to be found at sea and around the coasts, and both are caused by the air cooling until the dew point is reached.

Advection Fog

This is the true sea fog caused by warm moist air moving over cold water. The cold water cools the lowest layer of air to such an extent that the water vapour in the air condenses to form droplets of liquid water. This is known as the dew point.

Gust front

Fig 27   Circulation in a thunderstorm

When the wind comes before the rain,
soon we can set sail again.

Around the UK, advection fog is most common in Spring or early Summer when the water is cold and warm moist air from the southwest moves over it. There must be light winds to move the air over the water but once the wind speed reaches around 15 knots, the fog lifts off the water and forms low stratus clouds (*Fig 28*).

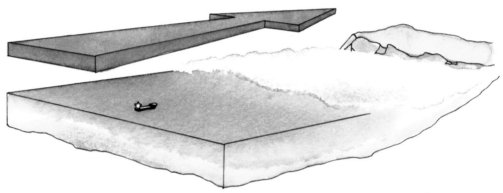

Fig 28   Warm, moist air passing over cold sea produces advection fog

This type of fog can last for long periods of time, often until there is a change in the air mass. There are some well-documented fog banks that form quickly when certain conditions prevail. The best known are the fog banks over the Grand Banks off Newfoundland, where cold water is brought south by the Labrador Current and warm moist air moves north with the Gulf Stream.

Closer to home the fog banks of eastern Scotland and northeast England are known as 'Haar' or 'sea fret', and are common in Spring and early Summer.

Air with a dew point higher than the sea water temperature is likely to give widespread fog. If it is similar, the fog will be patchy as the temperature of the sea varies a little with the change of tides and upwelling. Measuring dew point is not something that we regularly do on board a yacht but we can find reports on the internet from buoys and station circles.

Radiation fog

This is a land fog formed during clear nights under high pressure. Without any cloud cover to blanket the ground, the heat radiates upwards and is lost to space. As the ground cools, the air above it cools. If the air temperature falls below the dew point, fog is formed (*Fig 29* overleaf).

Although this is a land fog, it drifts down rivers and estuaries and finds its way over coastal waters. As the fog drifts out over the relatively warm sea, the air is warmed and starts to rise, mixing with the air at higher levels. This combination of warming and mixing is the beginning of the break up of the fog. The sun will warm the fog from above so over the open sea, radiation fog is quickly dispersed.

Fig 29   Radiation fog drifting down to the sea

The coldest part of the night is around dawn, and the hour after dawn can see the worst of the radiation fog. The heating of the sun will, on all but the worst days, quickly burn the fog off and a strong wind will rapidly disperse it. Tied up in a marina or at anchor we may be fog-bound in the early morning whilst out to sea it is clear.

Radiation fog is often worst in the autumn and winter, when on land it is reluctant to shift all day. The cold water of rivers lowers the local sea temperature, allowing the fog to spread to coastal waters, particularly in areas such as the Thames Estuary and eastern Channel.

Land forecasts are useful in picking up the chance of fog, as it is so important for road users. However, any night with clear skies and light winds is likely to produce radiation fog.

Knowing the type of fog is important for deciding when it is likely to clear; advection fog may well last until there is a change in wind direction and weather system, whilst radiation fog needs heat to burn it off and is likely to be only near the coast and in rivers and estuaries.

Waves

Waves are caused mainly by the wind, so their size depends mainly on the strength of the wind, on how long it has been blowing, and on its fetch - the uninterrupted distance that it has been blowing over water.

Over deep, open water, waves are likely to correspond reasonably well with the descriptions given in the Beaufort Scale. In coastal waters, however, the tidal stream and depth of water play an important part, particularly affecting the shape and steepness of the waves.

If the tidal stream is against the wind, the wave length (the distance between crests) will shorten by as much as 50%. The wave height, however, will stay much the same, so the

Wave height(H)
Wave length(L)
Wave steepness

Idealised wave

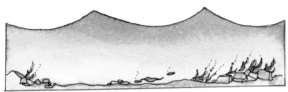

As waves grow they take on a more peaked shape

Actual wave shapes from a wave recorder showing little symmetry

Fig 30

steepness of each wave will increase significantly, producing what is described as a 'short, steep, sea'.

When the average slope of a wave increases to about 1 in 7 (a height/length ratio of about 0.14) it is likely to break. This dissipates its energy, and may leave a short-lived patch of relatively calm water. While it is breaking, however, a wave is at its most dangerous: big breakers are easily capable of capsizing or pitch-poling yachts, or of smashing windows or superstructures. The 1998 Sydney to Hobart Race turned into a disaster because the wind, blowing against a strong ocean current, produced breaking waves large enough to cause knockdowns and inversions which claimed six lives.

Shallow water has a similar effect: when a wave runs into water that is shallower than about half its wavelength, the water movement in the lower levels of the wave is restricted by the sea bed, so the whole wave has to slow down, shortening its wavelength, and making it steeper and more likely to break. This is why you often find waves breaking on a beach, and why harbours with shallow entrances or bars need to be treated with respect in onshore winds.

The effect will be particularly pronounced where headlands or narrow channels increase the speed of the tidal stream, or where a shallow and uneven sea-bed produces overfalls.

Squeeze zones

Where high pressure and low pressure move close together, we can get a zone where the isobars are squeezed together giving stronger winds. This can

Wind over tide in the Needles Channel

happen unexpectantly, if you're looking at a series of synoptic charts at 24 hour intervals: at each time step the low and high may be well separated, but for the low to have got to the second position it will have passed close to the high.

This is most likely to happen on the north side of high pressure in the Northern Hemisphere (south side in the Southern), where transient lows predominate.

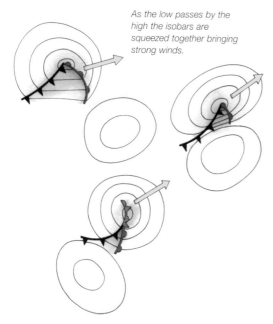

As the low passes by the high the isobars are squeezed together bringing strong winds.

Fig 31

Lee troughing

Wherever mountains interrupt the flow of wind there is likely to be an increase in pressure on the windward side and a drop in pressure on the leeward. This is a purely mechanical process but it can create a strong flow around the sides of the mountains. The Mistral is probably the best-known wind of this type but any range of mountains will have a similar effect on local wind conditions.

High mountains

Wind coming over high hills and mountain ranges will accelerate through passes and down valleys (*Fig 32*), in the lee of such mountains to give very strong, gusty winds.

Cold katabatic wind from high up the mountain is also a possibility at night.

These winds can be very localised making guides and pilot books an essential addition to cruising kit.

Katabatic winds can reach storm force in high latitudes near snow and ice covered mountains.

Wind chill

Wind chill is the name given to the cooling effect of wind. In hot weather, of course, a 'cooling breeze' is quite pleasant, though it can mask the early symptoms of developing sunburn. Early and late season sailors, however, need to be careful of the risk of developing hypothermia even if the actual air temperature is not particularly low.

An example:

With an air temperature of 10°C the wind chill equivalent temperature is 2.7°C in 20 knots of wind and minus 1.4°C in 40 knots.

Fig 32 Anchorages near mountains can give uncomfortable nights

High pressure and strong winds

Methodology  13

With all the information available, we can, and should be able to, improve on the forecast in the short term, and for the relatively small area that we sail in.

Putting it all together

The synoptic scale - dealing with large scale systems of high and low pressure.

The mesoscale - the scale between systems and individual clouds.

The shipping forecast areas.

The local or microscale - the small area where coastal features are significant and sea breezes are likely. It also includes the circulation around individual clouds. These are the conditions that we see and in which we sail.

The order of doing things:

- Study synoptic charts to get general conditions.

- Compare barometer, wind speed and direction, and clouds with those expected from the synoptic chart. Verify the forecast.

- Use the shipping forecast general synopsis to show how features will move, and the area forecasts to give the average wind for our sea area.

- Use the inshore waters forecast to see how the professionals think the general land mass will affect the wind.

- Refine the forecasts for our position within the sea areas for the movement of the synoptic systems.

- Look at the land and modify the forecast for land effects, both mechanical and thermal.

- Are there any weather dangers – squalls, thunderstorms, fog, etc? Keep a good weather eye, as there are always signs.

- Watch the clouds for local circulation.

- Check tides to avoid wind against tide conditions, especially around headlands or where the tidal streams or currents are strong.

- Continually monitor and log the wind, barometer, and clouds and compare them with the forecast.

By continually following the above we should be able to stay one step ahead of the weather, and by doing so have a safer and more enjoyable voyage.

New technology

Not many years ago, having a weather fax machine on board a yacht was considered the height of luxury, whilst the majority of us huddled by the radio listening out for 'Sailing By' and the shipping forecast.

New technology in weather forecasting

With modern communications and computers, we can turn a chart table into a mini Met Office, producing forecasts neatly overlaid on navigational charts. The internet has added a huge selection of additional information, whilst a dedicated aerial and black box will spew out satellite pictures from polar orbiting satellites to help clarify the situation.

The limitations are the speed of internet access and the cost of connecting to it by mobile phone or radio. The former is improving all the time although there can be a nasty shock when the bills arrive.

A lot of planning can be done ashore so we will look at what is available before going afloat and what we should be aiming for once sailing.

Just because a forecast can be graphically reproduced on a computer it does not mean that it is automatically accurate. Local land effects must still be taken into account and professionally prepared forecasts should be sought.

Ashore with full internet access

Any list of good weather sites will be out-of-date almost before it is published, but those given (see page 103) are ones that have been used and found reliable in the past.

The information available on the internet, can be split into two categories: forecasts and observations. The forecasts are from various computer models and not only show surface pressure, but give wind arrows, temperatures and other variables of varying usefulness.

The Met Office website gives access to shipping and inshore waters forecasts and, for a fee, access to planning and more detailed forecasts. It's also possible to receive forecasts by phone, fax, text message or even to talk personally to a forecaster.

For any planning, the synoptic charts are a must. These are readily available for up to 120 hours, or longer on the ECMWF site or from Top Karten.

A lot of the longer term charts do not show weather fronts and although they can be guessed at, it takes practise.

It must be remembered that the longer that the forecast is for, the less accurate it is likely to be and updates must be sought at every opportunity.

If we want to see wind arrows to get an indication of strengths and direction then the GFS model through Top Karten shows expected winds in steps of six hours extending to 180 hours. The area covered is large and we must look at likely local effects and also compare them with the synoptic charts. Meteo France also gives wind arrows that include the English Channel for 48 hours.

10m Wind (kt)

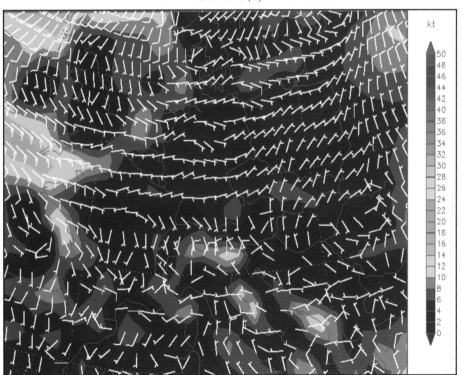

Wind arrows from American GFS via Top Karten

A word of caution – the wind arrows correspond to the synoptic chart from a particular model. The models will not always agree, therefore it is not a good idea to keep swapping between models.

Meteorological observations are particularly useful as they are fact and not open to interpretation. The Met Office provides hourly readings from a number of stations around the coast and it is also possible to download buoy reports which are the closest thing that we can get to real-time data about what is happening at sea.

Up-to-date analysis charts are also available on the web showing the station circles and the weather. These can be compared to the model charts to check how well the forecast is performing.

Satellite pictures add to the observations as they clearly show fronts and clouds. Cold fronts can be seen the most clearly with cumulus clouds behind. The analysis of satellite pictures on their own is difficult, but by comparing them to the synoptic charts it helps to bring the weather alive.

Keep surfing, as each day new sites are springing up. After a while, you will decide on a small group of sites that you use regularly and stick to. They are all raw data with no, or very little, forecaster input and must be used with caution alongside dedicated forecasts.

On board

Weather fax is one of the most useful sources of forecasts when away from the land. It is

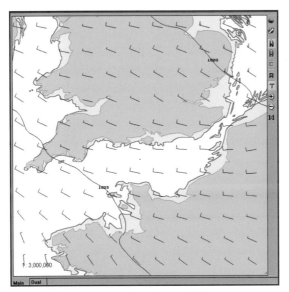

received through an SSB radio, either as a stand-alone unit, or connected to a laptop computer. Most of its output consists of synoptic charts, sometimes with wind speeds or wave heights added. The fax transmission follows a schedule given in ALRS volume 3, and in some almanacs, and is often updated on the internet.

Synoptic charts are one of the few things that are standard throughout the world so these are particularly useful when long-distance cruising. Be aware, though, some of the charts are transmitted for aviators so check that the one received is for sea level (MSLP).

Navigational chart with wind information overlaid. Courtesy of Weather Wizard

One of the most exciting developments is that met data can now be downloaded from the internet or email, and overlaid onto an electronic navigational chart. Some software will animate a series of forecasts covering several days, to help visualise the changing weather pattern in this navigational context. GRIB files, used by software such as MaxSea and Raytech, are produced free of charge by the American NOAA.

A similar, but not interchangeable system known as Weather Wizard uses information provided by the British Met Office.

In both cases, the files you get are part of a computer-generated forecast that covers the whole world, with a resolution of about 50km. The resolution is gradualy getting better, but

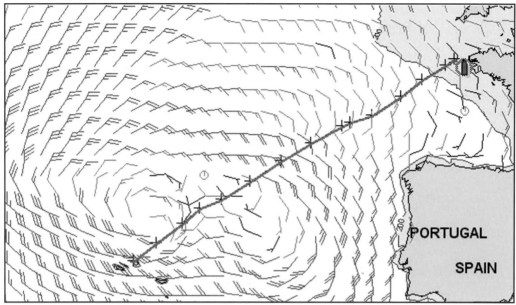

Diagram MaxSea routeing software with wind arrows and an optimum route for Brest to the Azores. Courtesy of MaxSea

at present, there is a risk that small scale features will be missed.

The absence of fronts drawn onto the chart is strange at first, although their positions can be estimated from the kinks in the isobars and/or the abrupt change in wind direction. Sailing through a front is likely to give a more abrupt wind change than the computer predicts.

The ability to enter the boat's sailing characteristics (polars), and to use a routeing programme that combines the weather forecast with the polars to produce an optimum course, is a very powerful routeing tool. How many of us have sat at the chart table wondering if it would be better to tack now or later? Or how much to crack off and where we will be when the expected wind swing arrives?

Check against reality

Whichever model you use, checking to see how it compares with reality is all-important. Whatever the computer may say it is the weather that is real, and once the model has strayed from reality it is unlikely to find its way back. It may be low tech, but the barometer is still one of the most important tools for the navigator. Combined with wind direction and strength it will tell you where you are within a system and gives an idea on how close to reality the model is.

There are other limitations too, particularly close to land where local conditions will modify the wind fields considerably.

Technology is continually changing and for a small cost personal graphical forecasts can be emailed directly to a boat at sea or to the crew ashore for planning before the start of a passage. The internet has many sites offering forecasts for planning up to ten days in advance as well as the opportunity to buy dedicated forecasts. All will need some user input and will, to some extent, need to be modified for local land effects. Mobile phones can also receive forecasts for specific areas.

Observations

Not only can these be found on the internet but some can also be received on WAP phones. Up-to-date observations help to verify forecasts, help in timing fronts, and can give advance warning of what is on its way. Supplementing forecasts, observations give a sense of reality to a forecast but it must be remembered that shore stations (particularly airports, where many of the observations originate) are subject to land effects so they may not give an accurate representation of the overall picture. Buoy reports, however, give a more representative report for an area.

Close to land, listening to the radio and getting the local forecast is still necessary as this will give a professionally prepared forecast using models with a much higher resolution, as well as a greater array of observations to give an accurate local picture. Each bit of information is just another piece in the jigsaw puzzle and just as no one should rely on one navigation system, neither should you rely on one piece of weather information.

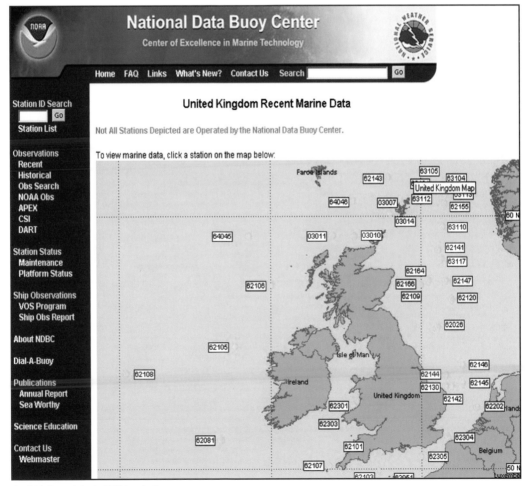

Available buoy data via NOAA website.

A few pointers

Visible images are at a better resolution than infra-red but are obviously limited to daylight hours. The infra-red measures the temperatures of the top of the clouds: the whiter the colour the colder the cloud top.

It's particularly useful to compare the satellite pictures with the synoptic charts. A site that does this for us is *Sembach.*

Sembach also publishes weather maps with fronts and expected cloud amounts and is well worth a visit.

Salellite picture with fronts overlaid.

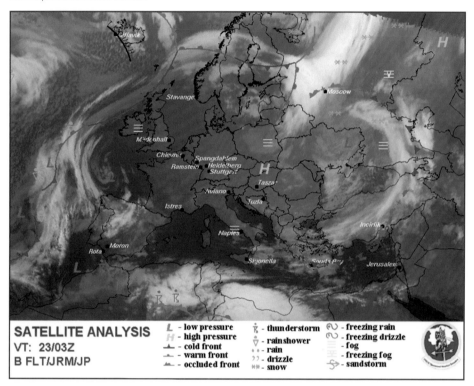

Hurricanes 15

Tropical revolving storms are intense depressions that, when they mature, generate winds in excess of Force 12 (64 knots) - sometimes up to 180 knots - accompanied by torrential rain. On land, they cause widespread damage and considerable loss of life, while at sea, the wind, waves, and rain take their toll on mariners.

Hurricanes are found around the world and in both hemispheres within a band between 5 and 20 degrees from the equator, and mainly during the summer months - though there are variations both in the names applied to them and to the exact season. In the Atlantic and eastern Pacific, for instance, they are known as hurricanes; in the Pacific they are called as typhoons; and in the Indian Ocean they are cyclones, while in Western Australia they are sometimes known as willy-willy. The main hurricane season is generally between July and October in the Northern Hemisphere and from December to March in the Southern, but the Bay of Bengal and Arabian Sea are different: there, the cyclone season is split, with cyclones most likely between April and June and between September and December.

Occasionally, however, hurricanes develop outside the usual season, so it is as well to be wary and to listen out for forecasts whenever you're sailing in tropical waters. Fortunately, the barometer, in such areas, tends to be generally steady, with just a regular daily rise and fall of a couple of millibars peaking at about 1000 and 2200 and reaching a minimum at about 0400 and 1600. Any pressure falling outside this normal daily range needs to be regarded as a possible early warning of a hurricane.

Four main conditions required for a hurricane to develop

1. More than five degrees away from the equator: this is because the Coriolis effect, which is essential to start the rotation of the storm, is zero at the equator itself.

2. The sea water temperature must be at least 26C° to a depth of at least 60m. In practice, this usually means the surface temperature must be nearer 28C°.

3. An unstable atmosphere allowing deep convection to great heights.

4. Some form of initial disturbance.

As a hurricane develops, it appears on satellite pictures as a distinctive circular patch of very dense cloud, with a clear 'eye' at the centre. Within the eye itself, winds are generally light but the sea is very rough and confused, because around it, in the 'eye wall' of the storm, winds may be blowing at up to 180 knots. Wind speeds generally decrease towards the outer edge of the storm: winds of 'hurricane force' (Force 12, or over 64 knots) are most likely to be within about 75 miles of the centre, while the wind further than 100 miles from the centre is unlikely to be more than gale force (Force 8, or 34-40 knots). A young hurricane usually moves fairly slowly, and in a generally westerly direction, angled slightly away from the equator, and gradually picking up speed until one of two things happens: either it reaches land, or it recurves - swinging abruptly towards the pole and then to the east.

Once a hurricane reaches land, it loses access to the massive reserves of heat energy that are stored in the sea, so it quickly dies. If it recurves, it soon finds itself over cooler water, so it either dies for the same reason, or else becomes weaker and more diffuse, and turns into a relatively ordinary mid-latitude depression.

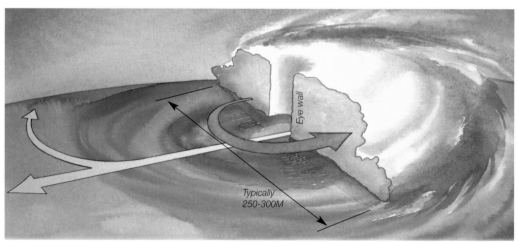

The structure of a Northern Hemisphere hurricane showing the likely tracts and cloud structure.

The effect of wind spiralling round a moving centre divides any hurricane into what are traditionally referred to as a 'navigable semicircle' and a 'dangerous semicircle'. It's easy to see why if you look at the Northern Hemisphere hurricane in the upper half of Fig 33 opposite. In the northern 'dangerous semicircle' the wind is flowing eastwards around the centre at perhaps 100 knots, but the whole thing is also moving eastwards at perhaps 20 knots, so the surface wind speed is actually 100+20=120 knots. In the southern 'navigable semicircle' by contrast, the circulating wind is flowing westwards at 100 knots, but it is partly offset by the fact that the hurricane itself is moving eastwards, to give a relatively (!) low surface wind speed of 100-20=80 knots.

Anywhere in the navigable semicircle of a Northern Hemisphere hurricane, broad reaching on starboard tack will take you away from the centre and away from any possible recurvature. In the Southern Hemisphere, broad reaching on port tack will have the same effect.

How to remember the Caribbean hurricane season

June – too soon.   July – stand by.

August – come she must.   September – remember.

October – all over.

However, hurricanes and tropical storms do not always know the rules and can happen outside the recognised season.

In the dangerous semicircle of a Northern Hemisphere hurricane, port tack is definitely not an option: it will lead you straight towards the worst conditions. Starboard tack is little better: unless the hurricane has already past you, it is only likely to prolong the agony by taking you parallel to the hurricane's track. If you are lucky enough to be able to make progress to windward under such conditions, you may just be able to escape to the north - where you will be ready to be caught by the same hurricane when it recurves!

Buys Ballot's Law (see page 28) is fundamental in deciding where you are in relation to the centre of the hurricane, because if you stand with your back to the wind, in the Northern Hemisphere, the centre will be to your left. In most hurricane areas, there are also very good hurricane forecasts, which will give enough warning of a hurricane and its expected track to give you time to either get out of its way or at least get across to the less dangerous side.

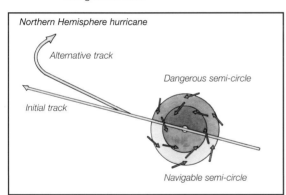

Northern Hemisphere hurricane
Alternative track
Dangerous semi-circle
Initial track
Navigable semi-circle

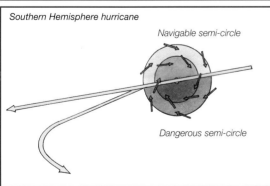

Southern Hemisphere hurricane
Navigable semi-circle
Dangerous semi-circle

Having been caught in a hurricane, it was not clear-cut as to what the tactics should have been. Caught in the dangerous semi-circle, the option of trying to cross the hurricane's track before it arrived was not appealing, but neither was staying where we were. In the end, the hurricane continued to curve to the east and passed some 70 miles to the north of our position, giving a very uncomfortable night with estimated wind speeds in excess of 90 knots for about 10 hours.

Torrential rain combined with the wind blowing the wave tops off prevented the boat from being overwhelmed, but at times it was touch and go. The overwhelming sensation though was one of noise; so intense that it became almost physical. Dawn finally came and the wind eased a little at which stage the waves were the most dangerous and sail was needed to get steerage-way.

Fig 33

Around the world

No book like this can cover all the possible weather patterns around the world, but we can look at the more famous (or infamous) winds likely to be met on an extended cruise.

Mediterranean winds

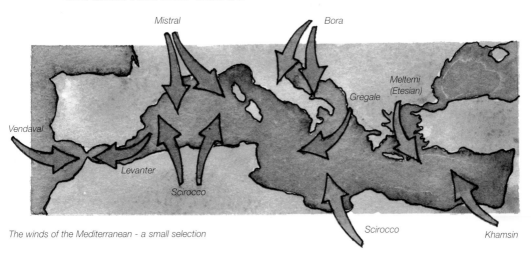

Mistral

Bora

Meltemi
(Etesian)

Gregale

Vendaval

Levanter

Scirocco

Scirocco

Khamsin

The winds of the Mediterranean - a small selection

The Mistral

One of the best known of the Mediterranean winds. The wind blows down the Rhône Valley with gale force ferocity and can affect a large percentage of the Western Mediterranean.

The Bora

A cold and dry katabatic wind from the mountains. In winter, it reaches storm force and is reputed to have overturned trains! It can develop with little warning although the local fishermen will tell you that there is a marked drop in the depth of water before it arrives.

Gregale

A strong northeast wind that is particularly important for sailors in Malta, where the main harbours are somewhat open to the northeast. Mainly a winter wind when there is high pressure to the north, and low to the south.

Meltemi or Etesian

A thermal wind from a heat low developing over Turkey and Asia. It is funnelled between Greece and Turkey and is strongest during summer afternoons.

Scirocco (Sirocco) or Khamsin

A southerly wind blowing off the deserts of North Africa, it has many local names. It is a hot and dry wind that carries a lot of sand with it. By the time it reaches the north Mediterranean coast it has picked up a considerable amount of moisture and may well be accompanied by stratus clouds and light rain. Hated by charter crews, it deposits fine orange sand over pristine yachts.

The Levanter and Vendaval

Easterly and westerly winds funnelled between southern Spain and Africa. The Levanter is generally moist and forms a banner cloud on the Rock of Gibraltar. If the wind is strong, violent eddies are likely around the Rock.

Other winds

The Southerly Buster

Rarely does a Sydney Hobart race go by without a Southerly Buster. It begins with a sudden change in wind direction from northwest (usually) to south or southeast behind a cold front, and is usually accompanied by a sudden fall in temperature. The fall in temperature can be as much as 20°C. If the rise in pressure is quick, the southerly wind can be gale force and it is often marked by a crescent-shaped roll cloud and accompanied by thunder.

Local winds
Everywhere has its local winds and pilot books are essential for gaining information about these winds. However by looking at the topography, it is possible to get a good idea of where strong winds are likely to be met. Local knowledge is important, and a good excuse for visiting the bar of the local yacht club!

Pampero

The Pampero is similar to the Southerly Buster and found off the coast of Argentina and Uruguay. A line squall of high intensity, it separates warm and cold air at the rear of a depression. A cigar-shaped roll cloud is seen on the approach of a Pampero accompanied by violent gusts of wind.

Harmattan

An east or northeast wind found in northwest Africa. It is a dry hot wind, and coming from the Sahara, it can bring large quantities of sand out to sea, severely restricting visibility.

Trade winds

The steady winds found in both hemispheres on the eastern and equator sides of the subtropical high-pressure belts. They vary in position and strength year on year but rarely fail completely. Moderate steady wind and sunshine have made sailing in the trades almost legendary. With a full moon and spinnaker set there can be few places that offer better sailing.

Towards the western and equator sides of the trade wind belt there is an increase of moisture in the air, giving rise to more squally weather. Details of the extent and strength of trade winds can be found on routeing charts.

Easterly waves

Sometimes found in the trade winds, easterly waves are shallow troughs moving westwards at between 10 and 20 knots. There is generally fair weather ahead of the trough and cloudy and squally weather behind. The wind follows the arrows shown in the diagram.

An easterly wave ⩔ indicates showers

Doldrums – ITCZ

This is the thermal equator of the world - moving north and south following the sun, but lagging by a couple of months. The mean position, however, is north of the equator. Over land the movement is much greater than over the sea bringing rainy seasons to tropical lands.

It is a band where the northeast and southeast trade winds meet – hence the name Inter-Tropical Convergence Zone (ITCZ). As with all surface convergence, this gives a general rising of air, lower surface pressure, and rain, usually in the form of huge squalls.

The weather in the ITCZ is therefore usually one of calm, interrupted by vicious squalls and towering black clouds. The gusts can reach gale force with stinging rain.

Doldrums

It is not always like this; sometimes we have an easy transition from northeast to southeast trade winds with few squalls, but at other times a wide band of variables is met. It has been likened to a pot of water boiling and not knowing where the next bubble is likely to break out.

Satellite pictures and computer models help in determining where the best crossing place is likely to be. The barometer is not much help as the changes in pressure are small. The Coriolis force, that spins the air around systems, is too small at the equator to have effect, so the surface air tends to travel direct from high to low pressure.

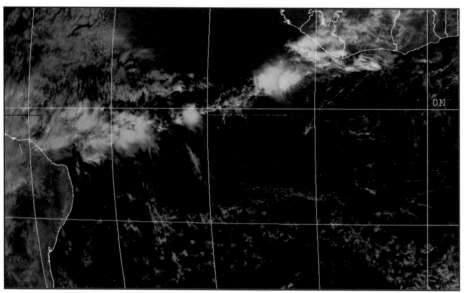

The ITCZ clearly seen from a satellite picture. This picture was for mid-Northern Hemisphere in winter when the ITCZ is at a southerly point.

Southern Hemisphere

Most of our examples throughout the book have been aimed at the Northern Hemisphere, but when we move south of the equator many things reverse.

Cape Horn - on a good day. Probably the most famous cape in the world.

In particular, the wind around high and low pressure is reversed in the Southern Hemisphere. That is, the wind around low pressure will circulate clockwise in the Southern Hemisphere and anti-clockwise around high pressure.

The surface wind will still head in towards the low pressure and out from the high, so that the surface wind is veered from the isobars.

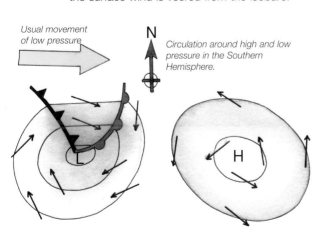

Usual movement of low pressure

N

Circulation around high and low pressure in the Southern Hemisphere.

So, in the Southern Hemisphere, the greater the surface drag, the more the surface wind is veered.

Gusts are therefore likely to be backed by a few degrees, and lulls veered.

The sea breeze direction will also change; in the Northern Hemisphere, a true sea breeze veers as the afternoon progresses, but in the Southern, it will back. It may be difficult to reverse our thinking, but it is really only a problem for people sailing from one hemisphere to the other.

The equator is a band of low pressure and weather systems do not travel across it, but rather we sail out of one system and into another.

The Roaring Forties

There are many stories from the days of the clipper ships to the modern round-the-world

races about the Roaring Forties and the Furious Fifties. These terms refer to the weather found in these latitudes in the Southern Hemisphere.

The Southern Ocean is unique as the only ocean that girdles the world with no land to disrupt the flow of water and depressions. Strong winds and high waves are the result and although the average wind direction is westerly in the Roaring Forties, in actuality the wind changes in direction quite significantly.

This is because there is a flow of deep depressions that arrive from the west. Like the depressions in the Northern Hemisphere they are made up of fronts with warm and cold sectors: the main difference is that the wind around these Southern Hemisphere lows travels in a clockwise direction. It is the seemingly continual flow of these depressions that gives the area such a fearsome reputation. Between the lows are ridges of high pressure that may last for a few hours or a couple of days.

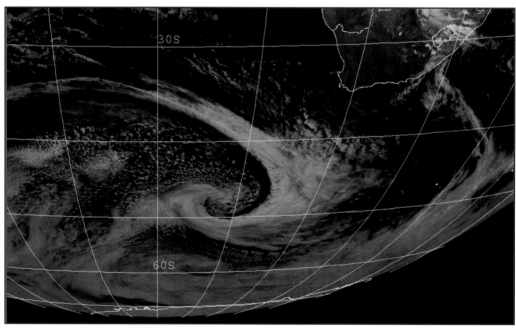

Southern Ocean storm captured by satellite

Ten top websites

http://www.metoffice.com

Inshore waters forecast, shipping forecast and a whole lot more. Good for observations from around the coast. In the aviation section, the ballooning forecast gives the chance of sea breezes developing.

http://www.metbrief.com

Really the only site you need as it has numerous links to all the major weather sites; including charts, satellite pictures and observations.

http://www.wetterzentrale.de/topkarten/

Lots of weather models including the familiar Met Office charts. Long range forecasts for up to 15 days can be useful for planning but only for a general idea.

http://www.xcweather.co.uk

Up-to-date observations from around the UK that can also be accessed by WAP phone. Observations also include buoy reports so very useful when sailing around the coast.

http://www.ndbc.noaa.gov/Maps/United_Kingdom.shtml

Buoy reports for around the British Isles

http://www.greatweather.co.uk

A huge number of links.

http://wwwghcc.msfc.nasa.gov/GOES

Interactive satellite pictures for anywhere in the world.

http://www.meteo.fr

Meteo France wind maps include the English Channel.

http://manati.wwb.noaa.gov/quikscat

Wind measured from satellites. Interesting to see what has been happening around the globe. Take care with the timing, as some of the data can be up to 22 hours old.

http://www.weatheronline.co.uk/forecast.htm

Information for around the world.

It is possible to spend many hours surfing for good sites although the information found is often the same, presented in different ways. Around the world, it is best to start with the local national met service as many have websites and links.

Glossary

A

Advection fog   Fog formed by warm moist air passing over a cold sea.

Air mass   Air with little change in temperature or humidity over a large horizontal distance.

Anabatic   A wind that blows up a slope, the opposite of katabatic.

Anemometer   Instrument for measuring wind.

Anticyclone   An area of high pressure.

Anticyclonic gloom   When pollution, fog and cloud is trapped by an inversion during an anticyclone.

Azores high   Semi-permanent high pressure that centres near the islands of the Azores

B

Backing   An anticlockwise swing in the wind direction.

Barometer   Instrument for measuring pressure.

Barograph   A recording barometer.

Beaufort scale   Wind force scale from 0 (calm) to 12 (hurricane).

Blocking high   High pressure that diverts the usual tracks of depressions.

Bomb   A very rapid fall in pressure as a low develops.

Boundary layer   The lowest level of the atmosphere where the surface has a big influence on the wind and weather.

C

Climate   Typical weather conditions based on observations and records over a considerable period of time.

Cloud streets   Lines of cumulus clouds running parallel to the wind.

Cold Front   The leading edge of a cold air mass replacing warmer air.

Condensation   When water vapour becomes a liquid.

Convection   The rising of heated surface air.

Convergence   A horizontal inflow of air.

Coriolis force   An apparent force caused by the spinning earth that deflects the wind to the right in the Northern Hemisphere and to the left in the Southern Hemisphere.

Cyclogenesis   The development of depressions.

Cyclone   Low pressure or depression.

D

Depression   Cyclone.

Dew point	The temperature to which air must be cooled to become saturated and at which fog forms.
Diurnal	Daily
Divergence	A horizontal outflow of air.
Doldrums	Low pressure area near the equator with generally light winds but heavy squalls.
Drizzle	Very small water droplets that fall slowly and reduce visibility.

E

Easterly wave	A disturbance in the trade winds that can occasionally turn into tropical storms.
Eddy	A departure from the main flow in either the air or water.
Eye	The clear centre of a hurricane surrounded by the eye wall.

F

Fetch	The distance the wind blows over open water.
Fog	Surface cloud with visibility less than 1,000m
Front	The zone between two air masses.
Frontogenesis	The formation or strengthening of a front.

G

Geostrophic wind	The wind blowing parallel to isobars.
Gradient wind	The wind blowing parallel to curved isobars.
Gulf Stream	A warm current that flows up the Eastern Seaboard of the USA before fanning out over the Atlantic towards Europe.
Gust	A short-lived increase in the wind.
Gust front	The leading edge of a violent downdraft from a thunderstorm or large squall.

H

Haar or Sea fret	A local name for advection fog in eastern Scotland and parts of eastern England.
Halo	A ring round the sun or moon from the refraction of light by ice crystals.
Horse latitudes	Belts of light variable wind associated with subtropical anticyclones.
High	Anticyclone.
Humidity	A general term expressing the water content of the atmosphere.
Hurricane	An intense tropical storm in which wind speeds exceed 64 knots.

I

ITCZ	Intertropical convergence zone. Where the northeast trade winds of the Northern Hemisphere meet the southeast trade of the Southern Hemisphere. (See Doldrums)
Inversion	An increase in air temperature with height.
Isobars	A line on a weather map joining places of equal pressure.

J

Jet stream	Strong wind in a narrow band usually at around 30,000 feet.

K

Katabatic A wind blowing down a slope. Usually cold, it can be strong.

Knot One nautical mile per hour.

L

Land breeze Thermally driven wind blowing from the land to the sea at night and the opposite to a sea breeze.

Lapse rate The rate at which the air usually cools with height.

Latent heat The heat absorbed or released as water changes state by evaporation or condensation.

Lee troughing A trough of low pressure forming in the lee of mountains.

M

Mackerel sky Cirrocumulus or altocumulus that looks a little like the scales of a mackerel and is usually a warning of an approaching depression.

Mare's Tails High wispy cirrus clouds.

Mesoscale The scale between large systems and localised weather.

Millibar (mb) A measure of atmospheric pressure.

Mist Poor visibility but better than 1 kilometre.

N

Nocturnal jet A strong narrow band of wind sometimes found above the boundary layer at night when an inversion has occurred.

O

Occluded front When the cold front overtakes the warm front.

Okta Used to measure cloud cover in eighths.

P

Polar front A front dividing warm tropical air from cold polar air.

Precipitation Any form of water; rain, hail or snow, that falls to the ground.

Pressure The weight of the atmosphere at any given point. Expressed in millibars. Standard pressure is 1013.2mb.

Pressure gradient The rate of change in pressure over distance. The closer the isobars the greater the pressure gradient and the stronger the wind.

R

Radiation fog Produced when the cooling of the land drops the air temperature to below its dew point. A land fog that can drift out to sea.

Roaring Forties Strong westerly winds found between 40 and 50 degrees south.

Roll cloud An elongated cloud sometimes found with a gust front or squall.

S

Scud Ragged fragments of low cloud moving rapidly below rain clouds.

Sea breeze A thermally produced onshore wind during the late morning and afternoon.

Secondary low A low pressure area that develops near an existing low.

Shower Precipitation from convective clouds.

Stability	The tendency for air to rise.
Squeeze zone	An area between high and low pressure where the isobars are squeezed together and there is strong wind.
Synoptic Charts	Weather charts covering a wide geographical area showing large systems.

T

Triple point	Where the cold, warm, and occluded fronts join.
Tropopause	The top of the troposphere.
Troposphere	The layer of the atmosphere from the surface to about 12km (depending on latitude).
Trough	A mini-front with generally cloudy and showery weather.

V

Veering	A clockwise swing in the wind direction.
Virga or fallstreaks	Precipitation falling from clouds that does not reach the surface.
Warm front	The leading edge of a warm air mass replacing a cooler air mass.
Wind	The movement of air in relationship to the earth's surface.
Wind sheer	The change in direction and speed of the wind with height.

A note about Units

Meteorology has a mixture of units that have developed over the years. Different branches have their favourite units and although research uses the more scientific SI units practical meteorologists have stuck with the more traditional.

This has led to an odd mixture of units. It is not uncommon to have cloud heights described in feet or thousands of feet, visibility in metres, and wind speed in knots - all within one forecast!

Pressure is generally given in millibars (mb) or hectopascal (hPa); 1mb is equal to 1hPa so these units are interchangeable. Older barometers may still show pressure in inches or millimetres of mercury.

> ### Conversion to millibars or hectopascal
>
> To convert to mb multiply:
> inches of mercury by 33.86
> mm of mercury by 1.33

The knot is the most common unit for wind speed but some countries, particularly Scandinavia, are likely to give wind speed in metres per second. For all practical purposes, the conversion is to multiply metres per second by 2 to get knots as:

1 knot = 0.514 metres/second.

Index

Page numbers in italics refer to a diagram or illustration. Subheadings are arranged in ascending page order.

TRAINING COURSES

140,000 people improve their skills every year

○ **Inland Waters**

○ **VHF Short Range Certificate**

○ **Basic Sea Survival**

Competent Crew
RYA Day Skipper, RYA/MCA Coastal Skipper
RYA/MCA Yachtmaster Offshore
RYA/MCA Yachtmaster Ocean
navigation and seamanship,
theory and practical

○ **Windsurfing**
beginners to wave sailing

○ **First Aid**

○ **Race Management**

○ **Powerboat and Personal Watercraft**

○ **Radar**

○ **Diesel Engine Maintenance**

○ **RYA National Sailing Scheme**

RYA

more boating knowledge from

www.rya.org.uk

For a training brochure, advice or details of personal membership call the RYA on 0845 345 0400 or email info@rya.org.uk

expert
knowledge
and
advice
online

RYA Shop

RYA

www.rya.org.uk

RYA Membership

Promoting and Protecting Boating

The RYA is the national organisation which represents the interests of everyone who goes boating for pleasure.

The greater the membership, the louder our voice when it comes to protecting members' interests.

Apply for membership today, and support the RYA, to help the RYA support you.

Benefits of Membership

- Access to expert advice on all aspects of boating from legal wrangles to training matters
- Special members' discounts on a range of products and services including boat insurance, books, videos and class certificates
- Free issue of certificates of competence, increasingly asked for by everyone from overseas governments to holiday companies, insurance underwriters to boat hirers

- Access to the wide range of RYA publications, including the quarterly magazine
- Third Party insurance for windsurfing members
- Free Internet access with RYA-Online
- Special discounts on AA membership
- Regular offers in RYA Magazine
- ...and much more

Join now - membership form opposite

Join online at **www.rya.org.uk**

Visit our website for information, advice, members' services and web shop.

Join the RYA now!

1 Important To help us comply with Data Protection legislation, please tick *either* Box A or Box B (you must tick Box A to ensure you receive the full benefits of RYA membership). The RYA will not pass your data to third parties.

☐ **A.** I wish to join the RYA and receive future information on member services, benefits (as listed in RYA Magazine and website) and offers.

☐ **B.** I wish to join the RYA but do not wish to receive future information on member services, benefits (as listed in RYA Magazine and website) and offers.

When completed, please send this form to: RYA, RYA House, Ensign Way, Hamble, Southampton, SO31 4YA

2

Title	Forename	Surname	Date of Birth			Male	Female
1.			D D / M M / Y Y			☐	☐
2.			D D / M M / Y Y			☐	☐
3.			D D / M M / Y Y			☐	☐
4.			D D / M M / Y Y			☐	☐

Address

Town _____ County _____ Post Code _____

Evening Telephone _____ Daytime Telephone _____

email _____ Signature: _____ Date: _____

3 Type of membership required: *(Tick Box)*

☐ **Personal** *Current full annual rate £33 or £30 by Direct Debit*

☐ **Under 21** *Current full annual rate £11 (no reduction for Direct Debit)*

☐ **Family*** *Current full annual rate £50 or £47 by Direct Debit*

* Family Membership: 2 adults plus any under 21s all living at the same address

4 Please tick ONE box to show your main boating interest.

☐ Yacht Racing   ☐ Yacht Cruising

☐ Dinghy Racing   ☐ Dinghy Cruising

☐ Personal Watercraft   ☐ Inland Waterways

☐ Powerboat Racing   ☐ Windsurfing

☐ Motor Boating   ☐ Sportsboats and RIBs

Please see Direct Debit form overleaf

Instructions to your Bank or Building Society to pay by Direct Debit

Please complete this form and return it to:
Royal Yachting Association, RYA House, Ensign Way, Hamble, Southampton, Hampshire SO31 4YA

Originators Identification Number

9	5	5	2	1	3

5. RYA Membership Number (For office use only)

To The Manager: Bank/Building Society

Address:

Post Code:

2. Name(s) of account holder(s)

3. Branch Sort Code

4. Bank or Building Society account number

Banks and Building Societies may not accept Direct Debit instructions for some types of account

6. Instruction to pay your Bank or Building Society

Please pay Royal Yachting Association Direct Debits from the account detailed in this instruction subject to the safeguards assured by The Direct Debit Guarantee.

I understand that this instruction may remain with the Royal Yachting Association and, if so, details will be passed electronically to my Bank/Building Society.

Signature(s)

Date

Office use / Centre Stamp

Cash, Cheque, Postal Order enclosed   £

Made payable to the Royal Yachting Association

Office use only: Membership Number Allocated

024